A.A. Dezin
Partial Differential Equations

Aleksei A. Dezin

Partial Differential Equations

An Introduction to a General Theory of Linear Boundary Value Problems

Translated from the Russian
by Ralph P. Boas

Springer-Verlag
Berlin Heidelberg New York
London Paris Tokyo

Aleksei A. Dezin
Steklov Mathematical Institute
ul. Vavilova 42, 117966 Moscow, USSR

Ralph P. Boas
Northwestern University
Evanston, IL 60201, USA

Title of the Russian original edition:
Obshchie veprosy teorii granichnykh zadach
Publisher Nauka, Moscow 1980

This volume is part of the *Springer Series in Soviet Mathematics*
Advisers: L.D. Faddeev (Leningrad), R.V. Gamkrelidze (Moscow)

Mathematics Subject Classification (1980):
35-01, 35F15, 35G15, 47B15

ISBN-13: 978-3-642-71336-1   e-ISBN-13: 978-3-642-71334-7
DOI: 10.1007/978-3-642-71334-7

Library of Congress Cataloging in Publication Data
Dezin, Aleksei A. Partial differential equations.
(Springer series in Soviet mathematics)
1. Boundary value problems. 2. Differential equations, Partial.
I. Title. II. Series.
QA379.D49 1987 515.3′53 87-9421

Preface

Let me begin by explaining the meaning of the title of this book. In essence, the book studies boundary value problems for linear partial differential equations in a finite domain in n-dimensional Euclidean space. The problem that is investigated is the question of the dependence of the nature of the solvability of a given equation on the way in which the boundary conditions are chosen, i.e. on the supplementary requirements which the solution is to satisfy on specified parts of the boundary.

The branch of mathematical analysis dealing with the study of boundary value problems for partial differential equations is often called mathematical physics.

Classical courses in this subject usually consider quite restricted classes of equations, for which the problems have an immediate physical context, or generalizations of such problems.

With the expanding domain of application of mathematical methods at the present time, there often arise problems connected with the study of partial differential equations that do not belong to any of the classical types. The elucidation of the correct formulation of these problems and the study of the specific properties of the solutions of similar equations are closely related to the study of questions of a general nature.

Among these are the following:

1. What accounts for the special position of the classical equations of mathematical physics (and their generalizations) among all possible equations?

2. Can one find a reasonable (in some sense of this term) boundary value problem for a randomly chosen equation, and if so, how?

3. What is the nature of the pathological phenomena that arise in the case of incorrectly posed boundary value problems?

These questions, and similar ones, need, of course, to be clarified, and are far from having complete answers. Nevertheless, it is clear that they should not be assumed to be merely speculative. The ability to orient one's self in unconventional situations is often valuable for a mathematician or physicist who is concerned with the solution of specific problems. For this reason, the author has tried to make the book accessible to the widest possible circle of readers.

Boundary value problems for partial differential equations constitute a rich and complicated subject, and can be considered from very diverse points of view. The basic approach in this book is through the theory of linear operators in Hilbert space. In certain constructions we also use spaces with other structures, but the Hilbert space of functions of integrable square is fundamental. In this connection, it is frequently most convenient to formulate the solvability properties of a boundary value problem in terms of the properties of the spectrum of an operator associated with the problem.

The first (introductory) chapter "Elements of spectral theory" is a brief exposition of the necessary facts from the corresponding parts of functional analysis.

In the second chapter we discuss general methods of associating a boundary value problem with a linear operator on Hilbert space.

The generality of the questions enumerated above makes it necessary to impose a number of quite stringent restrictions on the operators that we shall study. The elucidation of correct formulations of problems and the study of particular properties of their solutions for "nonclassical" equations is conveniently begun by the consideration of idealized models, for example by considering equations with constant coefficients, with part of the boundary conditions replaced by the condition of periodicity. This allows the application of some version of the method of separation of variables. In essence, the main part of the book (Chapters IV–VI) is based on the use of methods of this kind. By means of these we are led to the consideration of special classes of operator equations for which it is possible to obtain meaningful and rather complete results.

The reader can obtain additional details about the content of the book by looking through it. Numerous general remarks are contained in the introductory subsections, numbered "0".

In conclusion, I offer the following additional remarks. If the books in which the methods of functional analysis are applied to the study of boundary value problems are conditionally divided into two groups:

1) treatises on functional analysis in which differential operators are studied as concrete examples;

2) treatises on the theory of partial differential equations in which functional analysis is one of the methods employed;
then, putting this monograph into the second group, I would emphasize that my intention is that the basic theme should be an exposition of the mechanism of applying the general concepts of functional analysis to the study of definite classes of specific classical entities.

In conclusion, I take this opportunity to thank Professor Sh.A. Alimov for reading the manuscript and making many valuable comments.

A.A. Dezin

Preface to the English Edition

The main theme of this book is the study of how the solvability of a given linear partial differential equation depends on the choice of the boundary conditions; the principal methods are those of functional analysis. I feel that this theme deserves more attention than it usually receives. Rather than proving many general theorems, I have presented numerous special cases, for which more or less complete results are attainable, in order to illustrate various kinds of results. I hope that these examples will help the reader acquire enough intuition so that they can analyze the particular problems that arise in their own work. For a fuller discussion of the objectives of the book, the reader is referred to the preface to the Russian edition (above).

Shortly after the publication of the first edition, an approach was discovered to many of the problems that are discussed in the main part of the book; it is known as the model-operator method. It has become clear that with this approach one can analyze a large class of diverse problems, both from a unified point of view and in simplified formulations. A number of results in this direction are outlined in an appendix that contains brief summaries, kindly provided by Professor Boas, of some recent papers.

In conclusion, I want to express my gratitude to Professor Boas and to Springer-Verlag for producing this English edition, which should make the book accessible to a wider circle of readers.

Moscow, December 1986   A.A. Dezin

To the Reader

The book is divided into chapters; the chapters, into sections; the sections, into subsections. Formulas, theorems, and statements are numbered within each section. For a reference within a section, the number is given; for a reference to a different section of the same chapter, also the section (or section and subsection). Otherwise the chapter is also given.

Numbers in square brackets are references to the corresponding books or papers in the bibliography. A reference does not imply that the book or paper cited is the only (or principal) source of the information in question.

The "Halmos symbol" □ marking the end of a proof (possibly only an outline), or to emphasize its absence, is not used altogether systematically. In some cases where no confusion will result, it is omitted.

Definitions are not always set off in separate paragraphs. Frequently they are run into the text. Definitions of concepts are printed in italics.

Table of Contents

Chapter I
Elements of Spectral Theory

§0. Introductory Remarks

This chapter is introductory in nature. It contains the basic facts from the theory of linear operators that are fundamental for what follows. The contents of the chapter are more than amply covered by standard textbooks, for example those listed in the bibliography. Exact references are, in some cases, given in the text.

The reasons for including such a chapter in the book are evident: it is always convenient to have a brief summary of the information that is assumed to be known. Such a summary should eliminate the possibility of terminological discrepancies and serve the inexperienced reader as a compass in navigating through the ocean of propositions formed by the contents of the often terrifying volume of courses in functional analysis.

It is rather more difficult to justify the presence (or absence) of proofs. This is all the more true since the proofs that are given are sometimes quite detailed, whereas others are in the nature of hints. It is clear that the presence of proofs always gives a plan for more complete ones. Moreover, sometimes a proof lets us make a remark that seems to the author to be important; sometimes, it indicates a useful technical device; and sometimes the aim of a proof is simply to lighten the task of a reader who really wants everything to be proved.

Among the essential remarks mentioned above there belongs a mention of the details of our point of view, which is dictated by the fundamental subject of our study: the boundary value problem. Not all the facts enumerated in this chapter are used to the same extent. Some are presented only to complete the picture which will serve as general background to later constructions.

There are no examples in this chapter. The whole remainder of the discussion will serve as a set of examples for it.

§1. Basic Definitions

1.0. Introductory Remarks. The natural framework for the "abstract" spectral theory of operators, i.e. a theory that does not specify the way in which the operators are defined, is a complex Banach space. Although we shall be mainly concerned with a specific function space, namely Hilbert space, it is natural to present some facts in a more general setting. Moreover, it is just in this setting that they are presented in the standard treatises.

It should be noted that when one is concerned with "spectral theory" rather than "spectral theory of operators," contemporary references take as fundamental object an element of a certain Banach algebra. Hence the point of view of the present chapter may appear not to be abstract at all, but rather "concrete."

1.1. Fundamental Structure. If we start from the initial concepts of "naive set theory" – sets and relations, and follow the chain of axioms that enter into the definition of a Banach space, we obtain the following picture.

An *Abelian group* is a nonempty set G of elements $a, b, c, \ldots$, with a binary operation " $+$ " that associates with every pair a, b of elements of G a unique element $c \in G$ $(a+b=c)$. The operation " $+$ " is subject to the following additional requirements: it is associative $((a+b)+c=a+(b+c))$, commutative $(a+b=b+a)$; there is a neutral element 0 $(a+0=a)$; and for every $a \in G$ there is an inverse element $-a$ such that $a+(-a)=0$.

A *complex linear space* $\mathcal{K}$ is an Abelian group in which there is defined a multiplication of elements $a, b, c, \ldots$ by complex numbers $\alpha, \beta, \gamma, \ldots$, such that the following conditions are satisfied:

$$\alpha(a+b)=\alpha a+\alpha b, \qquad (\alpha+\beta)a=\alpha a+\beta a,$$
$$(\alpha\beta)a=\alpha(\beta a), \qquad\qquad 1a=a.$$

If we replace the complex numbers by the real numbers, we obtain the definition of a *real* linear space.

We emphasize that in restricting the class of *numbers* $\alpha, \beta, \gamma, \ldots$ in these definitions we are considering our objects from the point of view of *analysis*. An algebraist would have allowed the elements $\alpha, \beta, \gamma, \ldots$ in the definition to belong to an arbitrary field $\mathcal{F}$.

A *norm* is a nonnegative real function $\|a\|$, defined on elements $a \in \mathcal{K}$ and satisfying the following conditions:

1) $\|a\| = 0$ implies $a = 0$,
2) $\|\alpha a\| = |\alpha| \, \|a\|$,
3) $\|a+b\| \le \|a\| + \|b\|$.

A space $\mathcal{K}$ that has a norm is called a *normed linear space* (the qualifier "complex" or "real" will usually be omitted).

A sequence $\{x_n\}$ of elements of $\mathcal{K}$ is a *Cauchy sequence* if for every $\varepsilon > 0$ there is an integer $N(\varepsilon)$ such that the condition $m, n > N$ implies $\|x_n - x_m\| < \varepsilon$. A space $\mathcal{K}$ is complete if for every Cauchy sequence there is an element $x \in \mathcal{K}$ to which this sequence converges (in the ordinary sense).

A complete normed linear space is called a *Banach space* (*B*-space).

In a linear space the concept of linear dependence is defined in the usual way, and consequently so is the concept of *dimension:* the largest number of linearly independent elements. Although the spaces in which we are interested will usually be infinite-dimensional, we do not include infinite dimensionality in the definition of a *B*-space. Thus the set of complex numbers with the modulus as norm is an example of a one-dimensional *B*-space.

An incomplete normed linear space is called a *pre-Banach* space. Every pre-Banach space can be extended to a Banach space by the abstract process of adjoining the limits of the convergent sequences ([12], Chap. II, § 3.4).

Every *B*-space is simultaneously both a metric space and a topological space, but this aspect is without interest for our purposes.

A complex linear space $\mathcal{K}$ is a *pre-Hilbert* space if to every ordered pair of elements a, b there is assigned a complex number (a, b), their scalar product, satisfying the following requirements:

1) $(a, a) \geq 0$ and $(a, a) = 0$ implies $a = 0$;
2) $(a, b) = \overline{(b, a)}$ (the bar denotes the complex conjugate);
3) $(a + b, c) = (a, c) + (b, c)$;
4) $(\alpha a, b) = \alpha(a, b)$.

If in a pre-Hilbert space we set $\|a\|^2 = (a, a)$, it follows immediately from the classical Bunyakovsky-Schwarz inequality

$$|(a, b)| \leq \|a\| \, \|b\|$$

that the function $\|a\|$ is a norm, and therefore a pre-Hilbert space is automatically a pre-Banach space.

The scalar product is continuous: $\lim_k (a_k, b) = (\lim_k a_k, b)$.

A complete normed pre-Hilbert space is called a *Hilbert space*. Every Hilbert space is a *B*-space. In order for it to be possible to introduce a scalar product that generates a norm in a Banach space, certain special requirements have to be satisfied ([21], Chap. I, § 5; in that book the term "pre-Hilbert space" has a quite different meaning).

1.2. Special Subsets. A subset $\mathcal{B}'$ of a Banach space $\mathcal{B}$ which is in turn a *B*-space with the norm induced by then norm of $\mathcal{B}$ is called a *subspace* of $\mathcal{B}$.

We are often led to encounter a subset $\mathcal{B}' \subset \mathcal{B}$ which is a linear subspace but does not satisfy the condition of completeness in the norm of $\mathcal{B}$. We call such a subset a *linear manifold*.

The simplest way to form a linear manifold in $\mathcal{B}$ is to consider the *linear span* of a given subset $\mathcal{M} \subset \mathcal{B}$, i.e. the set of all finite linear combinations of elements of $\mathcal{M}$. If $\mathcal{B}'$ also contains all limit elements, i.e. limits (in the norm of $\mathscr{B}$) of sequences of elements of $\mathscr{M}$, then the corresponding *closed* linear span is a subspace of $\mathscr{B}$. This distinction naturally occurs only when $\mathscr{M}$ has an infinite number of linearly independent elements.

A subset $\mathscr{Q} \subset \mathscr{B}$ is *dense* in $\mathscr{B}$ if its closure is $\mathscr{B}$. A set $\mathscr{M}$ of elements is *complete* in $\mathscr{B}$ if the linear span of $\mathscr{M}$ is dense. A complete set $\{e_k\}$ of elements of $\mathscr{B}$ (finite or countable) forms a *basis* if, in the representation of each element, $x = \sum_k x_k e_k$, the numbers x_k are uniquely determined. Although there are many important examples of B-spaces without countable bases, we shall not encounter them here.

Turning now to a Hilbert space of more particular interest to us, we first notice the following fundamental proposition.

Lemma (on orthogonal expansion). *Let $\mathscr{M}'$ be a linear manifold in the Hilbert space $\mathscr{H}$, and let $\mathscr{N}$ be the set of elements $\varphi \in \mathscr{H}$ such that $(\varphi, y) = 0$ for every $y \in \mathscr{M}'$. Then $\mathscr{N}$ is a subspace of $\mathscr{H}$, and each $x \in \mathscr{H}$ has a unique representation of the form*

$$x = x_{\mathscr{M}} \oplus x_{\mathscr{N}} \qquad (1)$$

where $x_{\mathscr{M}} \in \mathscr{M}$ (the closure of $\mathscr{M}'$) and $x_{\mathscr{N}} \in \mathscr{N}$.

Remark. The subspace $\mathscr{N}$ is called the *orthogonal complement* of $\mathscr{M}$ (or $\mathscr{M}'$), and (1) is the *orthogonal expansion* of x. The symbol $\oplus$ indicates this, and is also used in the notation $\mathscr{H} = \mathscr{M} \oplus \mathscr{N}$.

Proof of the lemma. We evidently need to consider only the case when $\mathscr{H}$ is infinite. In that case, that $\mathscr{N}$ is a subspace follows immediately from the properties of the scalar product.

If $x \in \mathscr{M}$, the proposition is trivial. Let $x \notin \mathscr{M}$ and let $\inf_{y \in \mathscr{M}} \|x - y_n\| = d$. Then there is a sequence $\{y_n\}$ such that $\|x - y\| = d_n \to d$ as $n \to \infty$. Let us show that the sequence $\{y_n\}$ converges, i.e. the infimum is attained for some element $y \in \mathscr{M}$. Using the definition of the norm in $\mathscr{H}$ as the square of the scalar product, we obtain

$$d_m^2 + d_n^2 = \|x - y_m\|^2 + \|x - y_n\|^2 = \tfrac{1}{2}(\|2x - y_m - y_n\|^2 + \|y_m - y_n\|^2). \qquad (2)$$

Since

$$2\left\|x - \frac{y_m + y_n}{2}\right\|^2 \geq 2d^2,$$

we have the inequality

$$d_n^2 + d_m^2 - 2d^2 \geq \tfrac{1}{2}\|y_m - y_n\|^2,$$

which shows that the sequence $\{y_n\}$ converges to some $y \in \mathscr{M}$.

Let us show that $x - y \in \mathscr{N}$. In fact, the function F of the real parameter t,

$$F(t) = \|x - y + t z\|^2,$$

must have, for an arbitrarily given element $z \in \mathcal{M}$, a minimum for $t=0$, i.e. $F'(0)=0$. Hence (considering the pair z, iz of vectors) we conclude that $(x-y, z)=0$ for every $z \in \mathcal{M}$. Setting $x_{\mathcal{M}}=y$, $x_{\mathcal{N}}=x-y$, we obtain (1). The uniqueness of the representation is evident. $\square$

The preceding proof can be instructively "geometrized" (in an especially intuitive way in a real Hilbert space). As is easily seen, the discussion remains valid if the space $\mathcal{M}$ is replaced by any closed convex set (one for which $y_1, y_2 \in \mathcal{M}$ implies $(y_1 + y_2)/2 \in \mathcal{M}$). The element $x-y$ is said to be the perpendicular from x to $\mathcal{M}$; the chain of inequalities (2) uses the classical relationship between a diagonal and a side of a parallelogram. The existence of this property is a characteristic property of the Hilbert norm mentioned above. The nontrivial verification of the existence of the element y for which the infimum is attained is a consequence of the infinite dimensionality.

From the lemma we at once obtain the following corollary.

Corollary. *A set $\mathcal{M} \subset \mathcal{H}$ is complete if and only if the equation $(y, x)=0$ for every $y \in \mathcal{M}$ implies $x=0$.* $\square$

We can apply the classical process of orthogonalization to any countable basis $\{\varphi_k\}$ in Hilbert space, and thereby obtain an *orthonormal basis* $\{e_k\}$ that satisfies the conditions $(e_k, e_j)=\delta_{kj}$ (the Kronecker symbol). Then the coefficients of the expansion $x=\sum_k x_k e_k$ of an element $x \in \mathcal{H}$ are determined by the equations $x_k=(x, e_k)$. An orthonormal system $\{e_k\}$ is a basis if and only if

$$\|x\|^2 = \sum_k |(x, e_k)|^2 \tag{3}$$

for every $x \in \mathcal{M}$.

Remark. By using the availability of a countable orthonormal basis in $\mathcal{H}$ we can obtain a shorter (but less instructive) proof of the lemma on orthogonal expansions.

If $\{\varphi_k\}$ is a basis in $\mathcal{M}$, there exists a uniquely determined system of elements $\{\psi_k\}$ such that $(\varphi_k, \psi_j)=\delta_{kj}$. The system $\{\psi_k\}$ is also a basis, and is said to be *biorthogonal* to $\{\varphi_k\}$. For a pair of conjugate biorthogonal bases the coefficients of the expansions $x=\sum_k x_k \varphi_k$ and $y=\sum_k y_k \psi_k$ are determined by the formulas $x_k=(x, \psi_k)$, $y_k=(y, \varphi_k)$.

A basis $\{\varphi_k\}$ in $\mathcal{H}$ is called a *Riesz basis* if there are constants c_1, c_2 such that, for every $x \in \mathcal{H}$ that is represented in the form $x=\sum_k x_k \varphi_k$, we have the inequalities

$$c_1 \sum_k |x_k|^2 \leq \|x\|^2 \leq c_2 \sum_k |x_k|^2. \tag{4}$$

Inequalities (4) serve as a replacement for (3) in cases when the latter is not available.

1.3. Operators. A function $\mathbf{T}$ defined on a set $\mathfrak{D}(\mathbf{T}) \subset \mathscr{B}_1$ and making each element $x \in \mathfrak{D}(\mathbf{T})$ correspond to a unique element $y = \mathbf{T}x$, $y \in \mathfrak{R}(\mathbf{T}) \subset \mathscr{B}_2$, where $\mathscr{B}_1$ and $\mathscr{B}_2$ are B-spaces, is usually called an *operator*. The sets $\mathfrak{D}(\mathbf{T})$ and $\mathfrak{R}(\mathbf{T})$ are called the *domain* and the *range* of $\mathbf{T}$.

We shall consider only *linear* operators $\mathbf{T}$, i.e. operators that satisfy

$$\mathbf{T}(\alpha x + \beta y) = \alpha \mathbf{T}x + \beta \mathbf{T}y \tag{5}$$

for all numbers α, β and elements x, $y \in \mathfrak{D}(\mathbf{T})$.

Besides $\mathfrak{D}(\mathbf{T})$ and $\mathfrak{R}(\mathbf{T})$, the most important set associated with $\mathbf{T}$ is $N(\mathbf{T}) = \mathrm{Ker}\,\mathbf{T}$, the *kernel* of $\mathbf{T}$, i.e. the set of $x \in \mathfrak{D}(\mathbf{T})$ such that $\mathbf{T}x = 0$. It follows at once from (5) that the sets $\mathfrak{D}$, $\mathfrak{R}$ and N are linear manifolds.

An operator $\mathbf{T}^{-1}: \mathscr{B}_2 \to \mathscr{B}_1$ (read, "acting from $\mathscr{B}_2$ to $\mathscr{B}_1$") is called an *inverse* of $\mathbf{T}$ if $\mathbf{T}^{-1}\mathbf{T} = E$ (the identity mapping) on $\mathfrak{D}(\mathbf{T})$. A necessary and sufficient condition for the existence of $\mathbf{T}^{-1}$ is evidently that $N(\mathbf{T}) = 0$ (an operator $\mathbf{T}^{-1}$ defined in this way is sometimes called a *left* inverse).

The *norm* of $\mathbf{T}$ is $\|\mathbf{T}\| = \sup_{x \in \mathfrak{D}(\mathbf{T})} (\|\mathbf{T}x\|_2 / \|x\|_1)$ (the norms of x and $\mathbf{T}x$ are the norms in $\mathscr{B}_1$ and $\mathscr{B}_2$, respectively). An operator $\mathbf{T}$ is *bounded* if its norm is finite ($\|\mathbf{T}\| < \infty$). The following fact is an important consequence of the linearity of $\mathbf{T}$.

Lemma. *The operator* $\mathbf{T}$ *is bounded if and only if it is continuous, i.e. if a sequence* $\{x_n\}$ *in* $\mathscr{B}_1$ *converges* $(x_n \to x)$ *then the sequence* $\mathbf{T}x_n$ *in* $\mathscr{B}_2$ *also converges* $(\mathbf{T}x_n \to \mathbf{T}x)$.

An unbounded linear operator $\mathbf{T}$, considered within the framework of normed linear spaces, is, in a certain sense, a pathological object: its definition is only "weakly compatible" with the fundamental structure, the norm. The difficulties connected with the study of operators generated by differentiation are closely related to the fact that in the most "convenient" function spaces this study inevitably leads to the consideration of unbounded operators. The basic method of overcoming these difficulties involves the use of the boundedness of the inverse of a given operator and the introduction of a notion of closure that is weaker than boundedness (continuity).

An operator $\mathbf{T}: \mathscr{B}_1 \to \mathscr{B}_2$ is *closed* if $x_n \to x$ and $\mathbf{T}x_n \to f$ imply that $x \in \mathfrak{D}(\mathbf{T})$ and $\mathbf{T}x = f$.

A bounded operator (extended, in a natural way, by continuity; see below) is always closed. The converse is in general false. Nevertheless we have the following proposition.

Theorem (Banach). *A closed operator whose domain is the whole space* $\mathscr{B}_1$ *is bounded.*

This theorem, which is one form of the closed graph theorem, has many aspects. The reader may reduce its statement to the form given above, in which we shall use it.

In reformulating boundary value problems in the language of operator theory, we inevitably have to use one or another generalization of the solution of the equation under consideration, a generalization obtained by extending the domain of the operator of differentiation. Let us give an abstract version of such a procedure.

An operator $\tilde{T}$ is an *extension* of the operator T, $T: \mathscr{B}_1 \to \mathscr{B}_2$, if $\mathfrak{D}(T) \subset \mathfrak{D}(\tilde{T})$ and both operators coincide on $\mathfrak{D}(T)$.

Standard examples of the use of extensions are the extension by continuity to the whole space of a bounded operator with dense domain, and the closure of a given operator T, i.e. the construction of the minimal closed extension $\tilde{T} \supset T$ (if there is such an extension). As we observed above, we shall make extensive use of the fact that operators generated by differentiation have closed extensions.

The last definition in this subsection applies only to Hilbert spaces $\mathscr{H}_1$ and $\mathscr{H}_2$. Let $T: \mathscr{H}_1 \to \mathscr{H}_2$ and let the element $y \in \mathscr{H}_2$ have the property that there is an element $h \in \mathscr{H}_1$ such that the equation

$$(T x, y)_2 = (x, h)_1$$

is satisfied for every $x \in \mathfrak{D}(T)$ (scalar products in $\mathscr{H}_2$ and $\mathscr{H}_1$, respectively). We then define the operator T^* (the *adjoint of* T) by $T^*: \mathscr{H}_2 \to \mathscr{H}_1$, by setting $y \in \mathfrak{D}(T^*)$, $T^* y = h$ (under the conditions specified above). This definition is consistent if and only if $\mathfrak{D}(T)$ is dense in $\mathscr{H}_1$. Indeed, this condition guarantees that h is uniquely determined; T^* is evidently linear, and $\mathfrak{D}(T^*)$ contains at least the zero element.

The following consequence of this definition is especially useful.

Proposition 1. *An operator T^* is closed if it is the adjoint of some operator* T.

In fact, if $y_n \to y$ and $T^* y_n = h_n \to h$, we may take limits in the equation $(T x, y_n) = (x, h_n)$ because of the continuity of the scalar product. $\square$

Another useful proposition is conveniently stated in algebraic language. We draw the picture (known as a diagram) where the horizontal lines indicate passage to the adjoint operator; and the vertical lines, to the inverse.

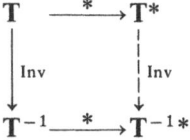

Proposition 2. *If, for a given operator $T: \mathscr{H}_1 \to \mathscr{H}_2$, we define the operations indicated on the diagram by the solid lines, then the dotted line makes the diagram commute, i.e. the operator T^{*-1} exists and $T^{*-1} = T^{-1*}$.*

Proof. The operator T^{*-1} exists. In fact, if the equation $(Tu, v) = (Tu, w)$ is satisfied for every $u \in \mathfrak{D}(T)$ then since $\mathfrak{R}(T)$ is dense in $\mathcal{H}_2$ (otherwise T^{*-1} would not exist) we have $v = w$, i.e. $N(T^*) = 0$.

Let us establish the inclusion $T^{*-1} \subset T^{-1*}$. Let $u \in \mathfrak{D}(T^{*-1})$, and let $u = T^* v$. Then for every $w \in \mathfrak{D}(T)$ we have the equation

$$(Tw, v) = (w, u). \tag{6}$$

If now $Tw = h$, $w = T^{-1}h$, then (6) yields

$$(T^{-1}h, u) = (h, v),$$

i.e. $u \in \mathfrak{D}(T^{-1*})$ and $T^{-1*}u = v = T^{*-1}u$.

The converse inclusion is verified by similar reasoning. □

1.4. Functionals. A linear operator $\mathscr{L} : \mathscr{B} \to \mathbb{C}$, where $\mathbb{C}$ is the Banach space of complex numbers ($\mathbb{C}$ may also be considered as a Hilbert space with the scalar product $(\alpha, \beta) = \alpha \bar{\beta}$) is called a *functional* (or *complex functional* to distinguish it from the *real* functionals $\mathscr{L} : \mathscr{B} \to \mathbb{R}$).

Since functionals are operators of a special kind, everything that we have said previously about operators carries over directly to functionals.

The set of all bounded functionals on a B-space $\mathscr{B}$ forms the *dual* space $\mathscr{B}^*$ of $\mathscr{B}$; this space plays a fundamental role in many situations. The special place occupied by the Hilbert space $\mathcal{H}$ among the Banach spaces is determined to a significant extent by the fact that $\mathcal{H}^*$ can be identified, in a natural way, with $\mathcal{H}$, i.e. in this sense we have selfadjointness. Let us prove the corresponding proposition.

Lemma (Riesz). *Let $\mathscr{L}$ be a bounded functional defined on a linear manifold $\mathcal{M}' \subset \mathcal{H}$. Then there is a unique element $h \in \mathcal{M}$ (the closure of $\mathcal{M}'$) such that*

$$\mathscr{L}(x) = (x, h) \tag{7}$$

for all $x \in \mathcal{M}'$. Moreover, $\|\mathscr{L}\| = \|h\|$.

Proof. We may consider $\mathcal{M}$ as a Hilbert space $\mathcal{H}_1 \subset \mathcal{H}$ (with scalar product given by the product in $\mathcal{H}$) and take $\mathscr{L}$ to be defined by continuity on the whole space $\mathcal{H}_1$.

If $\mathscr{L}(x) = 0$ for every $x \in \mathcal{H}_1$, we can set $h = 0$. If $\mathscr{L} \not\equiv 0$, then $N(\mathscr{L})$ (the kernel of $\mathscr{L}$) is a closed subspace different from $\mathcal{H}_1$. Consider the decomposition $\mathcal{H}_1 = N(\mathscr{L}) \oplus \mathfrak{Q}$. The subspace $\mathfrak{Q}$ is one-dimensional. In fact, for every pair of elements $x_1, x_2 \in \mathfrak{Q}$, with $\mathscr{L}(x_1) = \alpha_1 \neq 0$ and $\mathscr{L}(x_2) = \alpha_2 \neq 0$, we have

$$\mathscr{L}\left(\frac{x_1}{\alpha_1} - \frac{x_2}{\alpha_2}\right) = 0.$$

Let $q \in \mathscr{Q}$ be a basis element, $\|q\| = 1$, $\mathscr{L}(q) = \beta$. Then $\mathscr{L}(x) = (x, \bar{\beta} q)$ for every $x \in \mathscr{H}_1$. In fact, let us represent x in the form $x = x_N \oplus x_{\mathscr{Q}}$, and take $x_{\mathscr{Q}} = k q$. Then

$$\mathscr{L}(x) = \mathscr{L}(x_{\mathscr{Q}}) = \mathscr{L}(k q) = k \beta,$$

$$(x, \bar{\beta} q) = \beta(x, q) = k \beta(q, q) = k \beta.$$

It is evident that the element $h = \bar{\beta} q$ of $\mathscr{H}_1$ is unique. Moreover,

$$\|\mathscr{L}\| = \sup_{x \in \mathscr{H}_1} \frac{|\mathscr{L}(x)|}{\|x\|} \leq \frac{|\mathscr{L}(x)|}{\|x_{\mathscr{Q}}\|} = \frac{|k \beta(q, q)|}{|k|} = |\beta| = \|h\|,$$

and for $x \in \mathscr{Q}$ the inequality becomes an equality. □

The preceding proof is an obvious modification of the argument used for a finite-dimensional space $\mathscr{H}$. The equation $\mathscr{L}(x) = 0$ defines a hyperplane. It is, in general, infinite-dimensional, but its orthogonal complement is always one-dimensional. This fact was applied in the proof.

As we shall see, Riesz's lemma is a very useful tool in proving theorems about the existence of solutions of operator equations.

Corollary. *A bounded functional defined on a linear manifold $\mathscr{M}' \subset \mathscr{H}$ can be extended, with preservation of its norm, to the whole space $\mathscr{H}$.*

In fact, formula (6) evidently provides the required extension.

For a general B-space, the preceding corollary contains the content of the Hahn-Banach theorem. Its proof is considerably more complicated since in that case we do not have an explicit general form for linear functionals.

The Hahn-Banach theorem, together with Banach's theorem (above) and the uniform boundedness principle (which we shall not have occasion to use), otherwise known as the Banach-Steinhaus theorem, are the big theorems of the classical theory of B-spaces.

§ 2. The Spectrum of an Operator

2.0. Preliminary Remarks. One of the most useful features of the entities considered in linear functional analysis is the existence for many of them of analogies of a much simpler nature. The existence of these analogies enriches our intuition and has great heuristic value. These often originate in parallels with the theory of finite- and infinite-dimensional spaces (here it is appropriate to mention the book [8]), i.e. with analogies with vectors (points in Euclidean spaces) or functions (points in the Hilbert space of functions), and in parallels between the algebra of complex numbers and the algebra generated by the family of commutative operators on a given B-space.

Examples of the first type of analogy are in the remarks on the proof given above of the lemma on orthogonal decomposition and on the general form of functionals. In many questions of operator theory we find very fruitful applications of the second type of analogy. Here, in the first place, we have the comparison of objects: analytic functions with functions of an operator. The construction of a means of using such analogies is one of the fundamental problems of spectral theory.

For the construction of an algebra of operators, i.e. in carrying out arguments in which an essential role is played by the multiplication of pairs of operators T_1 and T_2, and all the more in defining functions of operators, we need to begin with the special case in which the operators act in the framework of a given B-space: $T_1, T_2: \mathscr{B} \to \mathscr{B}$; otherwise the construction is much more difficult.

Here it is appropriate to make a comment which is fundamental for understanding the basic purposes of this book. From the point of view of applications to the theory of boundary value problems, the restriction mentioned above is a very serious limitation on our circle of questions. A significant part of the problems connected with the left-hand side of a differential equation, considered as a linear operator L, relate, for example, to the question of a correctly chosen pair $\mathscr{B}_1, \mathscr{B}_2$ of spaces such that $\mathfrak{D}(L) = \mathscr{B}_1$, $\mathfrak{R}(L) = \mathscr{B}_2$, or to the determination of families of spaces $\mathscr{B}_1^\sigma, \mathscr{B}_2^\sigma$ such that $L: \mathscr{B}_1^\sigma \to \mathscr{B}_2^\sigma$ and the corresponding mapping is an isomorphism. This domain of applicability of the methods of functional analysis to the theory of boundary value problems is actually outside our field of interest.

Turning to the exposition of the basic concepts of spectral theory, we note that not only is its use important in the setting mentioned above, for constructing functions of an operator (in the first instance, of the differentiation operator), but also the corresponding language (terminology is given in 2.1) is very convenient for describing properties of specific operators that arise in the study of one or another class of boundary value problems for partial differential equations.

From the formal point of view the substance of spectral theory consists of the study of a special operator function of a complex parameter $\lambda \in \mathbb{C}$ associated with an operator $T: \mathscr{B} \to \mathscr{B}$. This function $T(\lambda): \mathscr{B} \to \mathscr{B}$ has the form

$$T(\lambda) \equiv T_\lambda = T - \lambda E, \tag{1}$$

where $E: \mathscr{B} \to \mathscr{B}$ is the identity operator. The presence of the operator factor E in (1) will usually not be explicitly indicated, i.e. the definition of T_λ will be written in the form

$$T_\lambda = T - \lambda.$$

We now enumerate the basic facts about the function T_λ and introduce the corresponding terminology; in the next subsection we shall attempt to clarify the fundamental role of this function in operational calculus, i.e. in the construction of functions of the operator T.

2.1. Basic Definitions. Let $T: \mathscr{B} \to \mathscr{B}$ be a given closed operator (in general, unbounded) with domain $\mathfrak{D}(T)$ which is dense in $\mathscr{B}$. Let T_λ be the operator function, with parameter $\lambda \in \mathbb{C}$, as defined above. We call a set $\rho(T)$ $\subset \mathbb{C}$ the *resolvent set* of T if, for every $\lambda \in \rho(T)$ the operator T_λ^{-1} exists, is bounded, and is defined on the whole space $\mathscr{B}$. We call the operator function $R_\lambda \equiv R_\lambda(T) = T_\lambda^{-1}$, with parameter λ, the *resolvent* of T.

Remark. The assumption that T is closed, which implies that T_λ^{-1} is closed (when the latter operator exists), makes it unnecessary to require that T_λ^{-1} is bounded if $\mathfrak{D}(T_\lambda^{-1}) = \mathscr{B}$. However, we prefer to mention this important property of the resolvent explicitly.

Proposition 1. *If* $\|T\| < 1$, *the point* $\lambda = 1$ *belongs to* $\rho(T)$.

A proof of this proposition can be given by considering the partial sums of the series $\sum_0^\infty T^k$ (the Neumann series), which provides a representation of the operator $(1 - T)^{-1}$ (since T is bounded, $\mathfrak{D}(T) = \mathscr{B}$ and the definition of arbitrary powers of T causes no difficulty). □

Proposition 2. *The set* $\rho(T)$ *is open in* $\mathbb{C}$.

In fact, if $\lambda_0 \in \rho(T)$, the operator

$$(T - \lambda_0)^{-1}[1 - \varepsilon(T - \lambda_0)^{-1}]^{-1}$$

is a bounded operator, with domain $\mathscr{B}$, provided that the modulus of ε is sufficiently small. At the same time,

$$(T - \lambda_0)^{-1}[1 - \varepsilon(T - \lambda_0)^{-1}]^{-1} = \{[1 - \varepsilon(T - \lambda_0)^{-1}](T - \lambda_0)\}^{-1}$$
$$= [T - (\lambda_0 + \varepsilon)]^{-1},$$

whence Proposition 2 follows. □

The closure of the set $\sigma(T) = \mathbb{C} \setminus \rho(T)$ is called the *spectrum* of the operator T. The point $\lambda \in \sigma(T)$ belongs to the *point spectrum* $P\sigma(T)$ of T if the operator T_λ^{-1} does not exist; the point $\lambda \in \sigma(T)$ belongs to the *continuous spectrum* $C\sigma(T)$ of T if T_λ^{-1} exists and the set $\mathfrak{D}(T_\lambda^{-1})$ is dense in $\mathscr{B}$, but T_λ^{-1} is unbounded; the point $\lambda \in \sigma(T)$ belongs to the *residual spectrum* $R\sigma(T)$ of T if T_λ^{-1} exists but the set $\mathfrak{D}(T_\lambda^{-1})$ is nondense in $\mathscr{B}$.

Evidently $\sigma(T) = P\sigma(T) \cup C\sigma(T) \cup R\sigma(T)$, and the sets on the right-hand side of the equation are disjoint.

If $\lambda \in P\sigma(T)$ then $N(T_\lambda) \neq 0$, i.e. the equation

$$(T - \lambda) u = 0 \tag{2}$$

has a nonzero solution. In this case λ is often called an *eigenvalue* of the operator T, and the nonzero solutions of equation (2) are the *eigenelements* (eigenvectors, eigenfunctions; see Chapter II), belonging to the corresponding eigenvalues.

This classification of the points of the spectrum is usually called the *coarse classification*. In fact, various refinements are possible. For example, for an eigenvalue λ the dimension of $N(T_\lambda)$ can be finite or infinite, for $\lambda \in R\,\sigma(T)$ the inverse operator can be bounded or unbounded, etc. Many texts on functional analysis deviate substantially from our terminology.

Proposition 3 (Hilbert identity). *If* $\lambda_1, \lambda_2 \in \rho(T)$ *then*

$$R_{\lambda_1} - R_{\lambda_2} = (\lambda_1 - \lambda_2) R_{\lambda_1} R_{\lambda_2}. \quad \square \tag{3}$$

From (3) there follows the commutability of the resolvents corresponding to the regular values of λ (those belonging to the resolvent set).

2.2. Functions of an Operator. As we have already remarked, the branch of spectral theory whose objective is the definition of functions of an operators is ordinarily called operational calculus. In the main part of the book the construction of an operational calculus will be developed either in a particularly simple situation – when a given operator has systems of eigenfunctions that form Riesz bases, or in a situation which makes use of special constructions that do not come within the scope of the elementary theory, but use some of its fundamental results.

The present subsection contains a sketch of some classical considerations that use the resolvent of an operator and the parallelism that exists between complex analytic functions and operator analytic functions. An exposition of the relevant questions, one very appropriate for our purposes, is given in the book [15], which unfortunately is not readily accessible.

In describing (and using) this parallelism, essential roles are played by the concept of holomorphy and the process of integration for operator functions. The ordinary concept of holomorphy involves single-valuedness and differentiability. The requirement of single-valuedness needs no comment, but the concept of differentiability can take various forms in connection with the possible introduction of different topologies. A convenient formulation of the concept of holomorphy that avoids these difficulties and implies differentiability in any reasonable sense (representability by series, etc.) is the following.

Definition. An operator function $S(z)$: $\mathscr{B} \to \mathscr{B}$ is *holomorphic* in $\mathscr{D} \subset \mathbb{C}$ if, for each bounded linear functional $\mathscr{L}$: $\mathscr{B} \to \mathbb{C}$, the corresponding complex function $\mathscr{L}[S(z)]$ is holomorphic.

Proposition. *If* $\lambda_0 \in \rho(T)$, *then* $R_\lambda(T)$ *is holomorphic in* λ *in some neighborhood of* λ_0. $\quad \square$

The most important property of holomorphic operator functions is that they possess an analog of Cauchy's integral formula. Let us describe the integration procedure that we need in order to write the formula.

Definition. The operator function $S(z)$: $\mathscr{B} \to \mathscr{B}$ of a complex variable $z \in \mathbb{C}$ is *continuous* at the point z_0 in the sense of the uniform operator topology if it is defined in a neighborhood of this point and for each $\varepsilon > 0$ there exists $\delta(\varepsilon) > 0$ such that $\|S(z) - S(z_0)\| < \varepsilon$ for every z in this neighborhood that satisfies the inequality $|z - z_0| < \delta$.

If $l = \{z(t), \ t \in [0,1]\}$ is a piecewise smooth curve in $\mathbb{C}$, and $S(z)$ is continuous (in the above sense) on l, and also $S(z_1) S(z_2) = S(z_2) S(z_1)$ for all z_1 and z_2 on l, it is then easy to define the operator

$$\mathbf{I}_l = \int_l S(z)\, dz, \qquad \mathbf{I}_l: \mathscr{B} \to \mathscr{B}.$$

It is sufficient to use the Riemann definition of the integral and consider the limit of sums of the form $\sum_i S(z_i')\, \Delta z_i$, where $z_i = z_i(t)$, $\Delta z_i = z_i - z_{i-1}$ and $0 = t_0 < t_1 < \ldots < t_n = 1$ is a suitable partition of the interval $[0,1]$. The limit is naturally taken under refinement of the partition and is independent of the parametrization of l. The holomorphy of the integrand implies that the integral is independent, in the usual sense, of the choice of the curve joining the given points.

We now formulate some fundamental facts that underlie the construction of an operational calculus.

Lemma. *Let* $\mathbf{T}: \mathscr{B} \to \mathscr{B}$ *be a bounded operator,* $\mathfrak{D}(\mathbf{T}) = \mathscr{B}$, $\|\mathbf{T}\| \le M$. *Then the integral*

$$(2\pi i)^{-1} \int_{|z| = 2M} (z - \mathbf{T})^{-1} dz \tag{4}$$

represents the identity operator $\mathbf{E}: \mathscr{B} \to \mathscr{B}$.

The proof of the lemma depends on the fact that under the hypotheses the spectrum of $\mathbf{T}$ is in the disk $|z| \le M$. The lemma is an analog of a proposition about the integral (4) in the case when $\mathbf{T} = \zeta \in \mathbb{C}$: the integral equals 1 or 0 according as the point ζ is inside or outside the disk $|z| \le 2M$.

By considering integrals of the form $(2\pi i)^{-1} \int_\Gamma (z - \mathbf{T})^{-1} dz$ over closed curves Γ that lie in $\rho(\mathbf{T})$ but surround only part of the set $\sigma(\mathbf{T})$, one can obtain "parts" of the operator $\mathbf{T}$, i.e. operators of the form $\mathbf{P}_\Gamma \mathbf{T} = \mathbf{T} \mathbf{P}_\Gamma$, where $\mathbf{P}_\Gamma$ is a projection (which commutes with $\mathbf{T}$) onto the corresponding subspace of $\mathscr{B}$.

Just as, in the theory of analytic functions, the integral (4), with $\mathbf{T} = \zeta \in \mathbb{C}$, is directly connected with Cauchy's integral formula

$$f(\zeta) = (2\pi i)^{-1} \int_\gamma f(z)(\zeta - z)^{-1} dz,$$

so the construction indicated above allows one to take the integral

$$(2\pi i)^{-1} \int_{|z| = 2M} f(z)(z - \mathbf{T})^{-1} dz \tag{5}$$

as the definition of the function $f(\mathbf{T})$. This definition is consistent with the direct definition of the simplest functions of $\mathbf{T}$ (for example, polynomials) and is quite suitable in other respects (see [5]). Here, it is understood, it is necessary to assume that f is analytic on the set $\sigma(\mathbf{T})$.

What has been said is sufficient to clarify the special role of the resolvent of an operator in the study of its structure.

These facts and their various corollaries and modifications are not at all trivial, but their pattern, unfortunately, must be made more complicated in order for us to pass to unbounded operators $\mathbf{T}$, for which the point spectrum inevitably contains the point $z = \infty$. The standard results in this direction are not adequate for the examples that are of interest to us. Some special constructions that are applicable in this case are discussed in Chapter VIII.

2.3. Connection Between the Spectra of an Operator and its Inverse. As is easily deduced from what was said above, the study of the spectrum of a bounded operator is essentially simpler than for an unbounded operator. We also saw above that the unboundedness of the operators generated by differentiation is partially compensated for, in many cases, by the boundedness of the associated inverse operators. Let us show that information about the spectrum of $\mathbf{T}^{-1}$ is very useful for the study of the spectrum of $\mathbf{T}$.

Lemma. *If* $\mathbf{T}: \mathcal{B} \to \mathcal{B}$ *is an unbounded closed operator for which* $\mathbf{T}^{-1}$ *exists, is bounded, and defined on the whole of* $\mathcal{B}$ ($0 \in \rho \mathbf{T}$)*, then a number* $\mu \neq 0$ *belongs to the spectrum of* $\mathbf{T}$ *if and only if* $\lambda = \mu^{-1}$ *belongs to the spectrum of* $\mathbf{T}^{-1}$.

Proof. It is enough to establish a correspondence between the resolvent sets of the corresponding operators. First let $\mu \in \rho \mathbf{T}$. Then the operator

$$\mathbf{T}(\mathbf{T}-\mu)^{-1} = \mathbf{E} + \mu(\mathbf{T}-\mu)^{-1}$$

is bounded. But $\mathbf{T}(\mathbf{T}-\mu)^{-1} = [(\mathbf{T}-\mu)\mathbf{T}^{-1}]^{-1} = \mu[\lambda - \mathbf{T}^{-1}]^{-1}$, i.e. $\lambda \in \rho \mathbf{T}^{-1}$.

On the other hand, if $\lambda \in \rho \mathbf{T}^{-1}$, the operator

$$\mathbf{T}^{-1}(\lambda - \mathbf{T}^{-1})^{-1} = [(\lambda - \mathbf{T}^{-1})\mathbf{T}]^{-1} = \lambda[\mathbf{T}-\mu]^{-1}$$

is bounded, and consequently $\mu \in \rho \mathbf{T}$. $\square$

It is clear that under our hypotheses we always have $0 \in C\sigma \mathbf{T}^{-1}$. It is also evident that if $\mu \in P\sigma \mathbf{T}$ then $\lambda \in P\sigma \mathbf{T}^{-1}$ and the dimensions of the corresponding eigensubspaces are equal.

§3. Special Classes of Operators

3.0. Preliminary Remarks. The operators generated by specific operations of analysis (differentiation, integration, multiplication by a function, etc.) have, as is easily seen, a whole array of special properties, for whose description we may use characteristics that can be defined abstractly. These characteristics are, as a rule, closely connected with properties of the spectra of the corresponding operators, and it is natural to present their definitions in a discussion of spectral theory.

The principal objects discussed in this section are the completely continuous (CC) operators. The study of the spectra of this class of operators can be carried rather far by starting from their basic property of being "almost finite-dimensional".

The last two subsections contain brief descriptions of various special kinds of operators (which are not CC operators).

3.1. CC Operators. Definition and Basic Properties. The inverses of operators generated by differentiation often have a stronger property than ordinary boundedness: they possess complete continuity (are CC operators). It is natural to express the initial definition in the language of mappings of $\mathscr{B}$-spaces.

Definition. A set $\mathscr{Q} \subset \mathscr{B}$ is *compact* in $\mathscr{B}$ if, from every infinite sequence $\{x_n\}$ of elements of $\mathscr{Q}$, it is possible to extract a subsequence that converges to an element x of $\mathscr{B}$.

The compactness of $\mathscr{Q}$ defined in this way is sometimes called *sequential* compactness. It follows from the definition that a compact set is closed and bounded.

Definition. An operator $\mathbf{T} \colon \mathscr{B}_1 \to \mathscr{B}_2$ is *completely continuous* if, for every bounded set $\mathscr{M} \subset \mathscr{B}_1$, the set $\overline{\mathbf{T}\mathscr{M}}$ (the closure in $\mathscr{B}_2$ of the image of $\mathscr{M}$ under the mapping $\mathbf{T}$) is compact in $\mathscr{B}_2$.

It follows immediately from the definition that a CC operator $\mathbf{T}$ is bounded (the image of the unit ball under $\mathbf{T}$ is compact and therefore bounded). The simplest example of a bounded operator that is not completely continuous is the identity operator (the unit ball in an infinite-dimensional space is not compact). A typical example of a CC operator is the operator of projection on a finite-dimensional subspace $\mathscr{B}' \subset \mathscr{B}_1$. That this example is typical is indicated by a theorem that asserts the "almost finite-dimensionality" of every CC operator.

Theorem (on approximation). *An operator* $\mathbf{T} \colon \mathscr{B}_1 \to \mathscr{B}_1$ *is completely continuous if and only if it admits uniform approximation by finite-dimensional operators.*

Remark. The proof of the second part of the theorem uses nonlinear approximations.

The sufficiency of the preceding criterion is established by using the easily verified fact that an operator $\mathbf{T}$ that is the limit of a uniformly convergent sequence $\{\mathbf{T}_n\}$ of CC operators ($\|\mathbf{T}_n - \mathbf{T}\| \to 0$ as $n \to \infty$) is completely continuous.

The proof of the necessity is also easily obtained if it is supposed to be known that, for a compact set $\mathcal{Q}=\overline{T\mathcal{M}}$ and every $\varepsilon>0$, there is a finite covering of $\mathcal{Q}$ by open balls $\{s_i\}_1^N$ of radius ε with centers at points $x_i\in\mathcal{Q}$ that generates a decomposition of unity: a system of functions $\{\varphi_i(x)\}$, $x\in\mathcal{Q}$, such that $\varphi_i\geq 0$, $\varphi_i(x)=0$ for $x\notin s_i$, and $\sum_i \varphi_i(x)=1$ for every $x\in Q$. Observing that T can be represented in the form

$$T y=\sum_i \varphi_i(T y)\,T y, \quad y\in\mathcal{M},$$

and forming the finite-dimensional transformation T_N:

$$T_N y=\sum_i \varphi_i(T y)\,x_i,$$

we see that

$$\|T y-T_N y\|=\sum_i \varphi_i(T y)\,\|T y-x_i\|\leq\varepsilon$$

uniformly in y, since $\varphi_i(T y)\neq 0$ only when $\|T y-x_i\|\leq\varepsilon$. $\square$

When considering spectral properties of CC operators, we must naturally suppose that $\mathcal{B}_1=\mathcal{B}_2=\mathcal{B}$. Here, as always, we are principally interested in the case of a Hilbert space $\mathcal{H}$.

Thus, now let T be a CC operator from $\mathcal{H}$ to $\mathcal{H}$. We notice a technical variant of the preceding theorem which is convenient in applications. Let $\{e_i\}_1^\infty$ be an orthonormal sequence. We say that $T\colon \mathcal{H}\to\mathcal{H}$ is *triangular* with respect to this sequence if, for each k,

$$T e_k=\sum_{i=1}^k \alpha_k^i e_i. \tag{1}$$

Theorem 1. *If T is a CC operator which is triangular with respect to the sequence $\{e_i\}$, then in (1) we have $|\alpha_k^k|\to 0$ as $k\to\infty$.*

The proof is easily given without any reference to the approximation theorem: Assuming that the theorem is false, we could construct a sequence $\{e_i'\}$ such that a convergent subsequence $\{T e_i'\}$ would not contain any convergent subsequence. $\square$

We also have the following proposition.

Theorem 2. *If $T\colon \mathcal{H}\to\mathcal{H}$ is a CC operator, so is T^*.*

Proof. We suppose it known that the adjoint of a bounded operator is bounded. If now $\{x_n\}$ is any bounded sequence, then, by considering the scalar square

$$(T^*[x_n-x_k], T^*[x_n-x_k])=(TT^*[x_n-x_k], [x_n-x_k])$$

and using the property that the composition of a CC operator and a bounded operator is again a CC operator, we see that we can select a convergent subsequence from $\{T^* x_n\}$.

3.2. CC Operators. Fredholm-Riesz Theory. The similarity, in the sense described above, between CC operators and finite-dimensional operators implies a similarity between equations involving these operators and finite systems of linear equations. The corresponding analogy, originally used in the theory of integral equations, can also be observed in abstract situations, and, generally speaking, for spaces much more general (see [35], [44]) than the Hilbert spaces that we have considered. The classical Fredholm theorems are consequences of some lemmas that specifically describe the properties of CC operators. We shall suppose, without going into details, that our CC operators act in a Hilbert space with a countable basis. The elements of this space will also be called vectors (especially in the term "eigenvector").

Lemma 1. *If* T *is a CC operator, then for each given* $\varepsilon > 0$ *there are only a finite number of linearly independent eigenvectors* u_n *of the operator* T *corresponding to the eigenvalues* λ_n *for which* $|\lambda_n| \geq \varepsilon$.

Proof. Suppose the contrary. Take an infinite sequence $\{u_n\}$ of linearly independent eigenvectors and construct an orthonormal sequence $\{e_n\}$:

$$e_k = \sum_1^k \beta_k^i u_i.$$

The operator T is triangular with respect to this orthonormal sequence. In addition (with obvious notation)

$$T e_k = \sum_1^k \beta_k^i \lambda_i u_i = \lambda_k e_k + \sum_1^{k-1} \beta_k^i (\lambda_i - \lambda_k) u_i = \lambda_k e_k + \sum_1^{k-1} \gamma^i e_i, \quad |\lambda_k| \geq \varepsilon,$$

which contradicts Theorem 1. $\square$

Corollary. *Let* T *be a CC operator. Then*
1. *To each nonzero eigenvalue of* T *there belong only a finite number of linearly independent eigenvectors.*
2. *The set of eigenvalues of* T *is at most countable.*
3. *Zero is the only possible limit point of the set of eigenvalues of* T.

Lemma 2. *If* T *is a CC operator then* $\Re(T_\lambda)$ *is, for every* $\lambda \neq 0$, *a closed subspace.*

Proof. Let $f \in \Re(T_\lambda)$. It is enough to establish the existence of a constant $c > 0$, independent of f, such that among the solutions of the equation $T_\lambda u = f$ there is a solution u for which $\|u\| \leq c \|f\|$.

Suppose that there is no such constant. Then there is a sequence

$\{f_k\} \in \Re(T_\lambda)$, with $\|f_k\| \to 0$ as $k \to \infty$, having the property that the solution of minimal norm of the equation $T_\lambda u_k = f_k$ satisfies $\|u_k\| = 1$. Take a subsequence $\{u'_k\}$ for which $\{T u'_k\}$ converges. From the equation

$$T u'_k - \lambda u'_k = f'_k \qquad (2)$$

it follows that the sequence $\{u'_k\}$ also converges. Taking limits in (2), we obtain $T u - \lambda u = 0$. But then $u'_k - u$ is a solution of (2), of arbitrarily small norm. Contradiction. □

Corollary. *If* **T** *is a CC operator then the only possible point of the continuous spectrum* $C \sigma T$ *is zero.*

Proposition. *If* **T** *is a CC operator then zero is a point of the spectrum of* **T**.

Proof. If this were not the case, the operator T^{-1} would exist, be bounded, and have domain $\mathcal{H}$; consequently $T^{-1} T = E$ would be a CC operator. □

Definition. A CC operator is called a *Volterra operator* (V-operator) if zero is the only point of its spectrum.

V-operators are a particularly important subclass of CC operators; they are connected with boundary value problems of a certain type.

Lemma 3. *If* **T** *is a CC operator and* $\lambda \neq 0$ *is an eigenvalue of* **T**, *then* $\Re(T_\lambda) \neq \mathcal{H}$.

Proof. Suppose the contrary. Starting from an eigenvalue $u_1 \neq 0$, we can construct an infinite chain $\{u_n\}$ of vectors determined by the equations $T_\lambda u_k = u_{k-1}$, $k = 2, 3, \ldots$. These vectors are linearly independent. We can use them to construct an orthonormal sequence $e_k = \sum_1^k \beta_k^i u_i$, with respect to which **T** is triangular. In the representation $T e_k = \sum_1^k \alpha_k^i e_i$ we have $\alpha_k^k = \lambda$, which contradicts Theorem 1. □

As an immediate corollary of this lemma we obtain Fredholm's first two theorems (however, other authors number them differently). Taking $\lambda \neq 0$ and writing the equation

$$(T_\lambda u, v) = (u, T_{\bar\lambda}^* v),$$

we see that, by Lemma 2, the orthogonal complement of the subspace $\Re(T_\lambda)$ is the same as the kernel of $T_{\bar\lambda}^*$, i.e. we have the decomposition

$$\mathcal{H} = \Re(T_\lambda) \oplus N(T_{\bar\lambda}^*) = \Re(T_{\bar\lambda}^*) \oplus N(T_\lambda),$$

so that Fredholm's first theorem is valid.

Theorem (Fredholm) **1.** *The equation*

$$\mathbf{T}_\lambda u = f, \quad \lambda \neq 0,$$

is solvable if and only if f is orthogonal to $N(\mathbf{T}_\lambda^*)$.

Now, using Lemma 3, we can state Fredholm's second theorem.

Theorem (Fredholm) **2.** *The number $\lambda \neq 0$ is an eigenvalue of $\mathbf{T}$ if and only if $\bar{\lambda}$ is an eigenvalue of $\mathbf{T}^*$.*

Fredholm's third theorem is somewhat different.

Theorem (Fredholm) **3.** *For every $\lambda \neq 0$ the dimensions $\dim N(\mathbf{T}_\lambda)$ and $\dim N(\mathbf{T}_{\bar{\lambda}}^*)$ are the same.*

Proof. Suppose the contrary, and let $e_1, \dots, e_k$ be an orthonormal basis for $N(\mathbf{T}_\lambda)$, and $\varepsilon_1, \dots, \varepsilon_k, \varepsilon_{k+1}, \dots$ an orthonormal basis for $N(\mathbf{T}_{\bar{\lambda}}^*)$. Construct a CC operator $\mathbf{W}$ by taking

$$\mathbf{W}u = \mathbf{T}u + \sum_1^k (u, e_j)\, \varepsilon_j.$$

The operator $\mathbf{W}$ has empty kernel. At the same time, ε_{k+1} is orthogonal to $\Re(\mathbf{W}_\lambda)$. Contradiction.

3.3. Selfadjoint CC Operators. As we noticed above, there is an important class of CC operators that have zero as the only point of the spectrum. Another important class consists of the CC operators that automatically have "enough" eigenvalues and eigenvectors; these are the selfadjoint CC operators. Finite-dimensional analogs of these classes are, on the one hand, the linear transformations representable in normal form as a single Jordan block, and on the other hand transformations represented by a symmetric (or Hermitian-symmetric) matrix.

We may remark that there is an extensive theory whose object is to obtain various normal forms of operators in the infinite-dimensional case, but we shall not be concerned with this.

As in the finite-dimensional case, the eigenvalues of a selfadjoint operator are evidently real, and the eigenvectors corresponding to different eigenvalues are orthogonal. The following geometric lemma can be used as a basis for the proof of the existence of nonzero eigenvalues of a selfadjoint CC operator.

Lemma 4. *Let $\mathbf{T}$ be a bounded selfadjoint operator. Then*

$$\|\mathbf{T}\| = \sup_x \frac{|(\mathbf{T}x, x)|}{\|x\|^2}. \tag{3}$$

Proof. Denote the right-hand side of (3) by S_T. Since evidently $\|T\|$
$=\sup\limits_{x,y} \dfrac{|(Tx, y)|}{\|x\| \|y\|}$, we have $\|T\| \geq S_T$. We must prove the opposite inequality.
We may suppose that $T \neq 0$. Let $x \neq 0$ be any element of $\mathscr{H}$ for which
$Tx \neq 0$. Let $\lambda > 0$ be determined by the equation $\lambda^2 = \|Tx\| \|x\|^{-1}$. Let
$u = \lambda^{-1} Tx$. Then

$$\|Tx\|^2 = (T\lambda x, u) = \tfrac{1}{4}\{(T[\lambda x + u], \lambda x + u) - (T[\lambda x - u], \lambda x - u)\}$$
$$\leq \tfrac{1}{4} S_T \{\|\lambda x + u\|^2 + \|\lambda x - u\|^2\} = \tfrac{1}{2} S_T \{\|\lambda x\|^2 + \|u\|^2\}.$$

But $\tfrac{1}{2}(\|\lambda x\|^2 + \|u\|^2) = \|x\| \|Tx\|$, i.e. $\|T\| \leq S_T$. $\square$

Lemma 5. *If* T *is a selfadjoint CC operator with* $\|T\| = |\lambda| \neq 0$, *then one of
the numbers* $\pm \lambda$ *is an eigenvalue of* T.

Proof. Let $\{x_n\}$ be a sequence with the property that $\|x_n\| = 1$ and
$(Tx_n, x_n) = \lambda_n \to \lambda$ as $n \to \infty$. Consider the sequence of numbers

$$\rho_n^2 = \|Tx_n - \lambda x_n\|^2 \geq 0.$$

We have
$$\rho_n^2 = (Tx_n, Tx_n) - 2\lambda(Tx_n, x_n) + \lambda^2 \leq 2\lambda^2 - 2\lambda \lambda_n, \tag{4}$$

and the right-hand side approaches zero as $n \to \infty$, i.e. $\rho_n^2 \to 0$. It follows that
if $\{Tx_n'\}$ is a convergent sequence, the subsequence $\{x_n'\}$ also converges,
$x_n' \to x$, $\|x\| = 1$. Passing to the limit in (4), we conclude that $Tx - \lambda x = 0$. $\square$

Theorem 3. *Let* T *be a selfadjoint CC operator,* $T \neq 0$. *Then there is a
finite or infinite sequence* $\{\lambda_n\}$ *of nonzero eigenvalues of* T, *which can be
indexed so that* $|\lambda_1| = \|T\|$, $|\lambda_k| \geq |\lambda_{k+1}|$. *If this sequence is infinite, then* $\lim\limits_{k} |\lambda_k|$
$= 0$, *and if* $\mathscr{M}$ *is the closed linear manifold spanned by the eigenvalues
associated with* $\{\lambda_k\}$ *then*

$$\mathscr{H} = \mathscr{M} \oplus N(T).$$

Proof. Let x_1 be an eigenvector associated with λ_1 and $\mathscr{H}_1$ the one-
dimensional subspace spanned by x_1. The decomposition $\mathscr{H} = \mathscr{H}_1 \oplus \mathscr{H}'$ reduces T, i.e. $T\mathscr{H}_1 \subset \mathscr{H}_1$, $T\mathscr{H}' \subset \mathscr{H}'$. The operator T' (the "part" of T acting
on $\mathscr{H}'$) is again a selfadjoint CC operator. If $T' = 0$, we are done; if $T' \neq 0$,
there exists an eigenvalue $\lambda_2 \neq 0$ of T', with $|\lambda_2| = \|T'\| \leq \|T\|$, and we may
repeat the preceding construction. The orthogonal complement of the linear
manifold $\mathscr{M}$ (defined in the statement of the theorem) necessarily belongs to
$N(T)$. $\square$

It follows from our discussion that the set of eigenvectors of a selfadjoint
CC operator can always be presented in the form of an orthonormal
sequence.

Theorem (Hilbert-Schmidt). *If* **T** *is a selfadjoint* CC *operator,* $\{e_k\}$ *is the orthonormal sequence of its eigenvectors, and* $f \in \Re(\mathbf{T})$, *there is a representation* $f = \sum_k f_k e_k$, *where* $f_k = (f, e_k)$. *If the sum is infinite, the equation is to be understood in the sense that the series converges in* $\mathcal{H}$.

Proof. The conclusion is evident if **T** is finite-dimensional. Otherwise let $\{e_k\}_1^N$ be the first (in the sense of Theorem 2) N eigenvectors of **T**, $f = \mathbf{T}u$, and

$$f = \sum_1^N f_k e_k + \rho_N, \qquad u = \sum_1^N u_k e_k + r_N.$$

Then $f_k = \lambda_k u_k$, $\rho_N = \mathbf{T} r_N$ and $\|\rho_N\| \le |\lambda_{N+1}| \, \|r_N\| \le |\lambda_{N+1}| \, \|u\|$, i.e. $\|\rho_N\| \to 0$ as $N \to \infty$. $\quad\square$

3.4. Selfadjoint, Normal, and Unitary Operators. The properties of the spectrum of a selfadjoint operator, reality of the eigenvalues and the existence of "enough" of them, are preserved if we relax the hypothesis of complete continuity. These properties permit significant modifications of the operational calculus introduced in §2. If we take $f(z) = z$ in formula (5) of §2, and use contours that shrink down to the real axis (see [15]) to obtain a system of projections we can represent a selfadjoint operator **T** as an integral

$$\mathbf{T} = \int_{-\infty}^{+\infty} \lambda \, d\mathbf{E}_\lambda(\mathbf{T}), \tag{5}$$

where λ is a real parameter, $\mathbf{E}_\lambda$ is a family of projections (known as a *resolution of the identity*) connected with **T**, and the action of the operator $d\mathbf{E}_\lambda(\mathbf{T}) \colon \mathcal{H} \to \mathcal{H}$ on the element u corresponds, roughly speaking, to projecting u on the eigenspace **T** corresponding to the eigenvalue λ.

If $\{\varphi_k\}$, the system of eigenfunctions of **T**, is a basis in $\mathcal{H}$, and $u = \sum u_k \varphi_k \in \mathfrak{D}(\mathbf{T})$, then (5) leads, as in subsection 3.3, to the formula

$$\mathbf{T}u = \sum \lambda_k u_k \varphi_k.$$

It is an extremely nontrivial fact that (5) continues to make sense even when the spectrum of **T** has a continuous part.

Functions of an operator **T** that has the representation (5) are defined by replacing λ by $f(\lambda)$ in (5).

Remark. The standard proof of (5) (see, for example, [1]) does not use the "general" operational calculus of §2, but is based on a different method.

The many advantages of (5) as compared with the formulas of §2 suggested the use of the representation of an arbitrary operator **T** as a sum of selfadjoint operators:

$$\mathbf{T} = \frac{\mathbf{T} + \mathbf{T}^*}{2} + i \frac{\mathbf{T} - \mathbf{T}^*}{2i}. \tag{6}$$

The realization of this idea encounters the difficulty that **T** and **T*** in general do not commute and the representation (6) does not yield a satisfactory simplification. However, it suggests the introduction of the class of operators **T** that can in fact be considered as sums $\mathbf{T} = \mathbf{T}_1 + i\mathbf{T}_2$, where $\mathbf{T}_1$ and $\mathbf{T}_2$ are selfadjoint. This class is characterized by the property called normality.

Definition. An operator **T** is *normal* if $\mathbf{TT^*} = \mathbf{T^*T}$.

The partial differential model operators, which will be very important in Chapter IV, will have exactly this property of being normal.

If we turn again to the analogy between the algebra of operators and the algebra of complex numbers, it is natural to compare the set of selfadjoint operators with the subalgebra of real numbers, and the normal operators with the complexification of this subalgebra. The analogy is at once both clear and incomplete: where do we put an "arbitrary" operator **T**?

The conventional style of exposition also obliges us to mention another class of operators, the analogs of the complex numbers of the form e^{ix} (x, a real number), for which $|e^{ix}| = 1$. These, the *unitary* operators, play an important role in physics, but we shall not use them later.

In a finite-dimensional Euclidean space, the unitary operators represent rotations. In a general Hilbert space $\mathcal{H}$, every unitary operator **U** admits a representation of the form $\mathbf{U} = \exp(i\mathbf{T})$, where **T** is selfadjoint. More precisely, there is an integral representation

$$\mathbf{U} = \int_0^{2\pi} e^{i\lambda} d\mathbf{E}_\lambda, \qquad \mathbf{E}_0 = 0, \qquad \mathbf{E}_{2\pi} = \mathbf{E},$$

where $\mathbf{E}_\lambda$ is a family of projections defined by a resolution of the identity, and generating (by formula (5)) a selfadjoint operator **T**.

3.5. Some Additional Conventions. In this subsection we define some terms that are not standard, but will be convenient in what follows. These refer to a class of operators that have not received any generally accepted names, but are often encountered in the study of the solvability of boundary value problems.

We shall call an operator $\mathbf{L}: \mathcal{H} \to \mathcal{H}$ with dense domain a *C-operator* if the operator $\mathbf{L}^{-1}$ exists and has the Volterra property (is a V-operator). The use of the letter C is suggested by the frequent occurrence of similar operators in the study of the Cauchy problem. It follows from the definition that a C-operator is an unbounded operator for which every finite point of the complex plane belongs to the resolvent set (zero belongs to the spectrum of $\mathbf{L}^{-1}$ and under our hypotheses cannot belong either to the point spectrum or the residual spectrum; the second part of this proposition follows from the relation between the spectra of $\mathbf{L}^{-1}$ and **L**).

It is convenient to call an operator L: $\mathcal{H} \to \mathcal{H}$ a $q\,C\text{-}operator$ (quasi-C-operator) if L has dense domain and the operator L^{-1} exists (but is not necessarily a CC operator) with zero the only point of its spectrum.

We call an operator L: $\mathcal{H} \to \mathcal{H}$ an M-$operator$ (M for "model") if it has a complete system of eigenfunctions that form a Riesz basis in $\mathcal{H}$. It is clear from what we have said in this chapter that the construction of an operational calculus is especially simple for M-operators.

If L^{-1} exists for an M-operator L, it is clearly again an M-operator. However, one of these operators may be bounded, and the other unbounded. At the same time, for example, the operator of multiplication by a constant is an M-operator (for which every element is an eigenelement).

Chapter II
Function Spaces and Operators Generated by Differentiation

§ 0. Introductory Remarks

This chapter discusses some results of a general nature connected with partial differential operators, considered in a bounded domain V of n-dimensional Euclidean space $\mathbb{R}^n$. We introduce some essential terminology and specify the nature of the problems that will concern us in what follows. The basic objective of the discussion is, in essence, the methods of associating specific objects of classical analysis with the abstract objects to which Chapter I was devoted.

In § 1 we introduce the Hilbert space $\mathbb{H}(V)$ of functions of square-integrable modulus; this is fundamental for what follows. Besides the formal definition, we make a number of supplementary comments whose object is to make the exposition more accessible for readers for whom the Riemann integral is more familiar than the Lebesgue integral. These comments will emphasize the aspects of Lebesgue theory that are essential for the constructions that we shall carry out.

The next two sections contain the definitions and preliminary studies of the basic concepts concerned with the "general theory" of boundary value problems. Further sections (to which an introduction is given at the beginning of § 4) discuss various aspects of the definition of the adjoint operator, considered in a concrete situation, and the construction of the special apparatus that allows us to obtain the necessary results. The corresponding constructions are of independent interest, and will be discussed in more detail than is required for the foundations of Chapters IV–VII.

§ 1. The Space $\mathbb{H}(V)$

Let V be a bounded domain in the Euclidean space $\mathbb{R}^n$, with boundary $\partial V = S$ consisting of a finite number of pieces of sufficiently smooth (for example, of class C^2) $(n-1)$-dimensional surfaces that intersect at nonzero angles. The following constructions can also be carried out under rather less restrictive hypotheses, but some regularity of ∂V is indispensable. The nature of the minimal requirements under which the results of this chapter remain valid is not of interest here.

The basic function space in which the subsequent discussion will be carried out is the Hilbert space $\mathscr{L}^2(V) \equiv \mathrm{IH}(V) \equiv \mathrm{IH}$ of complex functions on V with moduli of Lebesgue integrable square.

In defining the space $\mathrm{IH}(V)$, it will be convenient to argue in the following way. Let $C(V)$ be the set of functions that are continuous on the closed domain $\bar{V}$; these form a complex linear space with the ordinary operations of pointwise addition and multiplication by complex numbers. Defining a scalar product on $C(V)$ by the equations

$$(u, v) = \int_V u \bar{v} \, dV, \quad u = u(x), \quad v = v(x),$$

$$dV \equiv dx, \quad x \in V, \quad dx = dx_1 \dots dx_n, \tag{1}$$

makes $C(V)$ into a pre-Hilbert space. Completing this under the norm generated by the scalar product, we obtain the Hilbert space $\mathrm{IH}(V)$.

The investigation of the nature of the "ideal elements" adjoined to $C(V)$ as a result of this abstract process is a topic for the theory of functions. It is known that every element of the space $\mathrm{IH}(V)$ constructed in this way can be identified with a class of functions that are Lebesgue integrable and coincide almost everywhere.

If, taking the integral in (1) in the Riemann sense, we wish to remain in the framework of the Riemann theory of integration, we must take into account that not every element of $\mathrm{IH}(V)$ is represented by an integrable function. Nevertheless, for every element $u \in \mathrm{IH}(V)$ we can always define, for example, the number $\lim_{n \to \infty} (u_n, 1)$ which is independent of the choice of the Cauchy sequence $\{u_n\} \in C(V)$ that represents the element u, and can be written symbolically in the form $\int_V u \, dV$.

In any particular case, the following construction can be carried out without reference to Lebesgue theory, at the expense of supplementary limiting processes. However, making the corresponding effort throughout would be rather uneconomical.

An essential fact is that $\mathrm{IH}(V)$ is a complete (Hilbert) space, in which the linear manifold $C(V)$ is, by definition, dense. In discussing the embedding $C(V) \subset \mathrm{IH}(V)$ we keep the following remark in mind. The statement "the element u of IH is a continuous function" is to be interpreted in the sense that the corresponding class of functions contains a continuous function $u(x)$, $x \in V$ (evidently uniquely determined) with which this class can be identified in all discussions. Corresponding remarks hold also for embeddings of the linear manifolds $C^k(V)$ and $C^\infty(V)$ (the k times differentiable or infinitely differentiable functions) in $\mathrm{IH}(V)$.

Remark. It may be helpful for the reader to verify directly that the element $u \in \mathbb{H}(0, 1)$, represented by the function that equals 1 on $(0, \frac{1}{2})$ and zero on $(\frac{1}{2}, 1)$, has no representation by a continuous function.

In connection with the preceding remarks, in many cases (for example, in the following lemma) it is convenient, relying on results from the theory of functions, to speak of elements of $\mathbb{H}(V)$ as if they were functions (defined and finite almost everywhere). Later we shall need a property of such functions, often called "continuity in the mean".

We suppose that $u(x) \in \mathbb{H}(V)$ is defined on all of $\mathbb{R}^n$ by setting $u \equiv 0$ for $x \notin V$, and let

$$\mathscr{B}_\delta u = \sup_{|h| \le \delta} \{ \int_V |u(x+h) - u(x)|^2 \, dx \}^{\frac{1}{2}}.$$

Here

$$x + h = (x_1 + h_1, \ldots, x_n + h_n), \qquad |h|^2 = h_1^2 + \ldots + h_n^2.$$

Remark. In introducing this definition of $\mathscr{B}_\delta u$, we rely on the fact that a function $\tilde{u}(x)$ defined in a domain $\tilde{V} \supset V$, agreeing with $u \in \mathbb{H}(V)$ in V, and zero on $\tilde{V} \setminus V$, is an element of $\mathbb{H}(\tilde{V})$. We may suppose that this fact is known, but we could also prove it by constructing (starting from the corresponding sequence for $u(x)$) a sequence of functions $\{ \tilde{u}_i \}$ that are continuous in the closed region $\overline{V}$ and satisfy $\tilde{u}_i \to \tilde{u}$ in $\mathbb{H}$.

Lemma (on continuity in the mean). *For each given element* $u \in \mathbb{H}(V)$, *the number* $\mathscr{B}_\delta u$ *tends to zero as* $\delta \to 0$.

Proof. It is enough to discuss the behavior of $\mathscr{B}_\delta$ in, for example, $\delta \le 1$. Let $V \subset V_1$, where V_1 is a bounded domain in $\mathbb{R}^n$ such that $x + h \in V_1$ for all $x \in V$ when $|h| \le 1$. Let $\tilde{u}(x) \in \mathbb{H}(V_1)$ be a function that coincides with the given function $u(x)$ in V and is equal to zero on $V_1 \setminus V$ (see the remark above). Let $\{ \tilde{u}_i(x) \}$ be a sequence of continuous functions converging to $\tilde{u}(x)$. Since according to the definition of $\mathscr{B}_\delta u$ the extension of $u(x)$ outside V is zero, and $|h| \le 1$, we have

$$\{ \int_V |u(x+h) - u(x)|^2 \, dx \}^{\frac{1}{2}} = \{ \int_V |\tilde{u}(x+h) - \tilde{u}(x)|^2 \, dx \}^{\frac{1}{2}}$$

$$\le \{ \int_V |\tilde{u}(x+h) - \tilde{u}_i(x+h)|^2 \, dx \}^{\frac{1}{2}}$$

$$+ \{ \int_V |\tilde{u}_i(x+h) - \tilde{u}_i(x)|^2 \, dx \}^{\frac{1}{2}} + \{ \int_V |\tilde{u}_i(x) - \tilde{u}(x)|^2 \, dx \}^{\frac{1}{2}}.$$

With a given $\varepsilon > 0$, we may choose an integer i so that the first and third terms on the right-hand side each do not exceed $\varepsilon/3$, and then choose $\delta > 0$ so that

$$|\tilde{u}_i(x+h) - \tilde{u}_i(x)|^2 \le \varepsilon^2 \, (9 \text{ mes } V)^{-1}$$

for $|h| \le \delta$ (taking account of the uniform continuity of $u_i(x)$ in the closed domain $\overline{V}_1$). This makes $\mathscr{B}_\delta u < \varepsilon$ for the specified choice of δ.

§2. Differential Operations and the Maximal Operator

We may define, in the usual way, a *linear differential operation* on functions $u(x) \in C^m(V)$:

$$\mathbf{L}(D)u \equiv \sum_{|\alpha| \leq m} a_\alpha D^\alpha u. \tag{1}$$

Here $\alpha = (\alpha_1, \ldots, \alpha_n)$ is an integral-valued multi-index and

$$D^\alpha = D_1^{\alpha_1} \ldots D_n^{\alpha_n}, \quad D_k \equiv \frac{\partial}{\partial x_k}, \quad |\alpha| = \alpha_1 + \ldots + \alpha_n.$$

The coefficients a_α can be either complex numbers or complex functions $a_\alpha = a_\alpha(x)$ belonging at least to $C(V)$.

Let us explain the sense in which we use the term "operation". In the applications we may, of course, immediately consider $\mathbf{L}(D)$ as an operator $\mathbf{L}: \mathbb{H}(V) \to \mathbb{H}(V)$, defined on the linear manifold $C^m(V) \subset \mathbb{H}(V)$ (we again emphasize that both continuity and differentiability are always considered in the closed domain $\bar{V}$). However, in the first place, by an operator we shall always mean the closure in $\mathbb{H}$ of a given operation; and, in the second place, with a single operation – an expression of the form (1) – we shall as a rule connect a whole family of different operators from $\mathbb{H}(V)$ to $\mathbb{H}(V)$.

Let us demonstrate by the simplest example that an operation of the form (1), defined in the "classical" sense, is not closed in $\mathbb{H}$.

Example 1. Let $V = (0, 1)$, an interval of the real line, and consider the operation $\mathbf{L}(D) = D_x$ as an operator $D_x: \mathbb{H} \to \mathbb{H}$ with domain the linear manifold C^1 of all functions that are continuously differentiable on $[0, 1]$. The operator D_x so defined is not closed. In fact, if

$$u(x) = \begin{cases} x & \text{for } 0 \leq x \leq \frac{1}{2}, \\ 1 - x & \text{for } \frac{1}{2} \leq x \leq 1, \end{cases}$$

it is easy to construct a sequence of smooth functions $u_k \to u$ for $k \to \infty$ (convergence in $\mathbb{H}$) such that

$$D_x u_k \to f(x) = \begin{cases} 1, & 0 \leq x < \frac{1}{2}, \\ -1, & \frac{1}{2} < x \leq 1, \end{cases}$$

(convergence in $\mathbb{H}$), but nevertheless $u \notin C^1$, i.e. does not belong to $\mathfrak{D}(D_x)$.

Since we suppose the domain V to be given, in the future we shall not indicate it explicitly in the notation. When we consider the closure of an operation $\mathbf{L}(D)$ in $\mathbb{H}$ it is indifferent whether we suppose it to be originally defined on C^m or on C^∞, and ordinarily we select the latter.

If an operator $\mathbf{L}(D)$ is initially defined on C^∞, its closure in $\mathbb{H}$ is called the *maximal operator*, $\tilde{\mathbf{L}}: \mathbb{H} \to \mathbb{H}$, generated by $\mathbf{L}(D)$.

A detailed analysis of this definition is given in what follows. An element $u \in \mathbb{H}$ is assigned to $\mathfrak{D}(\tilde{\mathbf{L}})$ if there is a sequence $\{u_k\} \in C^\infty$ such that

$$u_k \to u, \qquad \mathbf{L}(D) u_k \to f \in \mathbb{H} \tag{2}$$

(convergence in $\mathbb{H}$)[1].

The term "maximal", as applied to the operator $\tilde{\mathbf{L}}$, indicates that among the operators $\mathbf{L}: \mathbb{H} \to \mathbb{H}$ connected in the ordinary way with the operation (2), the operator $\tilde{\mathbf{L}}$ has, in a definite sense, the largest domain $\mathfrak{D}(\tilde{\mathbf{L}})$. When considering the closure of an operator we must, of course, be certain of the consistency of the corresponding method. In other words, we must be certain that there cannot be, for the given operator $\mathbf{T}: \mathbb{H} \to \mathbb{H}$, two different sequences $u_k' \to u$ and $u_k'' \to u$ such that $\mathbf{T} u_k' \to f'$ and $\mathbf{T} u_k'' \to f''$ (convergence in $\mathbb{H}$), with $f' \neq f''$.

Example 2. Let $V = (0, 1)$, the interval on the real line, let the operator $\mathbf{T}: \mathbb{H} \to \mathbb{H}$ be defined so that $\mathfrak{D}(\mathbf{T}) = C \subset \mathbb{H}$, and for each element $u \in C$ let the element $f = \mathbf{T} u$ be defined by the equation $f(x) = \text{const} = u(\frac{1}{2})$. There evidently exist sequences $\{u_k'\}$, $\{u_k''\}$ of continuous functions that converge in $\mathbb{H}$ to, for example, the function $u_0(x) \equiv 1$, but such that $\mathbf{T} u_k' \equiv 0$, $\mathbf{T} u_k'' \equiv 2$ (for arbitrary k). Therefore this operator $\mathbf{T}$ does not have a closure.

It will be a consequence of the following discussion that under our hypotheses an operation $\mathbf{L}(D)$ always has a closure. We shall, however, not dwell on this fact here. It will be noticed in the appropriate place.

Throughout the following discussion the central role will be played by operations $\mathbf{L}(D)$ with constant coefficients a_α. In this case the kernel $N(\tilde{\mathbf{L}})$ of the maximal operator is different from zero when $m \geq 1$, i.e. the solution of the equation

$$\tilde{\mathbf{L}} u = f \tag{3}$$

(equation in $\mathbb{H}$; the operator $\tilde{\mathbf{L}}$ understood in connection with (2)) is not unique. The elements that belong to the kernel are, for example, functions of the form $\exp\left(\sum_1^k c_k x^k\right)$ for a corresponding choice of the constants c_k.

It is convenient to present this remark in the following form.

Proposition 1. *If $\mathbf{L}(D)$ is an operation with constant coefficients, $m \geq 1$, then the point spectrum of the corresponding operator $\tilde{\mathbf{L}}: \mathbb{H} \to \mathbb{H}$ fills the entire complex plane C.*

If we take the operation $\mathbf{L}(D)$ in (1) to be a classical operator (the Laplace operator; operators that appear in the wave equation or the equation of heat conduction), or their simplest generalizations, then equation (3) corresponds to the classical equations, considered in V without any sup-

---

[1] Another standard way of defining the closure is by introducing the graph of an operator (see [5]). The closure of the graph yields the closure of the operator.

plementary boundary conditions. Then, in addition to propositions on the nonuniqueness of the solutions, one can also assert the solvability of equation (3) for arbitrary right-hand sides $f \in \mathbb{H}$.

It turns out that with constant coefficients in the operation (1) the analogous proposition (on the solvability of (3) for an arbitrary $f \in \mathbb{H}$) remains valid also in the case of an arbitrary operation $\mathbf{L}(D)$. However, the proof of this is not trivial. We shall obtain it after a number of auxiliary constructions.

§ 3. The Minimal Operator and Proper Operators

3.1. The Minimal Operator. We denote by C_0^∞ the linear manifold of functions in C^∞ that satisfy the additional condition of reducing to zero together with their derivatives of all orders on S, the boundary of V.

From the point of view of the theory of boundary value problems, the "narrowest", or minimal, operator connected with the operation (1) can be obtained by choosing as initial domain for $\mathbf{L}(D)$ the manifold C_0^∞. More precisely, the *minimal operator* connected with the operation $\mathbf{L}(D)$ is the operator $\mathbf{L}_0$: $\mathbb{H} \to \mathbb{H}$ defined as the closure in $\mathbb{H}$ of the operation $\mathbf{L}(D)$ defined initially on C_0^∞.

In what follows, we shall not pause each time to describe in detail the closure method (or domain of the corresponding operator), which makes use of a limiting process similar to the one used in (2), § 2.

Our definitions immediately imply the inclusion $\mathbf{L}_0 \subset \tilde{\mathbf{L}}$, i.e. the maximal operator is an extension of the minimal operator.

If we now turn to operations with constant coefficients and argue in a way similar to what we did for equation (3) of § 2, it is natural to expect that the equation

$$\mathbf{L}_0 u = f \tag{1}$$

always has a unique solution, but this solution is far from existing for all right-hand sides $f \in \mathbb{H}$.

In fact, as we shall see, under our hypotheses these facts are exactly equivalent to the proposition on the solvability of equation (3) of § 2 for arbitrary $f \in \mathbb{H}$ and on the nonuniqueness of the corresponding solution.

These properties of equation (1), when stated in terms of spectral theory, imply the following proposition.

Proposition 1. *If* $\mathbf{L}(D)$ *is an operation with constant coefficients,* $m \geq 1$, *then the residual spectrum of the corresponding operator* $\mathbf{L}_0$: $\mathbb{H} \to \mathbb{H}$ *fills the whole plane* $\mathbb{C}$.

3.2. Transposed Operations and Adjoint Operators. The preceding remarks on the connection between the properties of equations (3) of § 2 and

(1) recall the classical Fredholm theory (subsection 3.2 of Chapter I), and it is not surprising that the proofs of the corresponding propositions are based on our attraction to the consideration of the operators adjoint to $\tilde{L}$ and L_0.

If the coefficients a_α of the operation $L(D)$ in (1) of §2 belong to class C^m, then the uniquely determined transpose (or formal adjoint of $L(D)$), the operation $L^t(D)$, is connected with $L(D)$ by the relation

$$(L(D)\, u, v) = (u, L^t(D)\, v),$$

which must be satisfied for all $u, v \in C_0^\infty$. The transition from $L(D)$ to $L^t(D)$ is based on integration by parts. This is always practicable for sufficiently smooth coefficients a_α.

At the same time, for the operator $\tilde{L}$ and L_0 connected with $L(D)$ we can define in the ordinary way (subsection 1.3 of Chapter I) the adjoint operators $\tilde{L}^*$, L_0^*: $\mathbb{H} \to \mathbb{H}$ (in the sense of operator theory; the domains of L and L_0 are evidently dense in $\mathbb{H}$). Here, in many important cases, we have the equations

$$\tilde{L} = (L_0^t)^*, \qquad L_0 = (\tilde{L}^t)^*, \tag{2}$$

$$\tilde{L}^t = L_0^*, \qquad L_0^t = \tilde{L}^*. \tag{3}$$

The first equation in (2) underlies the construction given in the following subsection. It will be proved in §6.

Meanwhile we make the following observation. The inclusion $\tilde{L} \subset (L_0^t)^*$ is evident. In fact, if $u \in \mathfrak{D}(\tilde{L})$, $v \in \mathfrak{D}(L_0^t)$, it is sufficient to write the equation

$$(L(D)\, u_i, v_k) = (u_i, L^t(D)\, v_k)$$

for the elements u_i, v_k of the approximating sequences that appear in the corresponding definitions, and take limits as $i, k \to \infty$. The proof of the inclusion $(L_0^t)^* \subset \tilde{L}$ is not trivial, and requires the use of regularization or of what are known as averaging operators.

The operator $(L_0^t)^*$ is sometimes called the weak extension of $\tilde{L}$, and the question of the validity of equations (2) and (3) is a special case of the problem of the equivalence of the weak and strong extensions of differential operations (see §4).

3.3. Existence of Proper Operators. Suppose that a differential operation $L(D)$, of the form (1), §2, is given in the domain V. Using the same notation as above, we introduce the following definition, which is fundamental for what follows.

Definition. An extension L: $\mathbb{H} \to \mathbb{H}$ of the operator L_0 ($L_0 \subset L$) is called *resolvable* if the operator L^{-1} exists and is defined on the whole of $\mathbb{H}$.

A restriction L: $\mathbb{H} \to \mathbb{H}$ of the operator $\tilde{L}$ ($L \subset \tilde{L}$) is called *proper* if the operator L^{-1} exists and is defined on the whole of $\mathbb{H}$.

A closed operator $L: \mathbb{H} \to \mathbb{H}$ which is simultaneously a resolvable extension of L_0 and a proper restriction of $\tilde{L}$ is called a *proper operator* generated by the operation $L(D)$.

The central problem considered in the basic chapters of the book is the problem of discovering methods for describing the proper operators that are generated by a general differential operator with constant coefficients, and the study of the dependence of the properties of these operators on the nature of the original operation and on conditions that determine the domain of the proper operator.

Remark. An operator $L: \mathbb{H} \to \mathbb{H}$ that we have called proper is often called a solvable extension generated by the operation $L(D)$. This terminology is inconvenient for detailed discussion of the corresponding circle of questions. We illustrate this by some examples.

Example 1. Let V be the interval $(0, b)$, $L(D) = D_x$, and let the operator $T: \mathbb{H} \to \mathbb{H}$ be defined on $\mathfrak{D}(\tilde{L})$ by the equation

$$T u = D_x u + \int_0^b \frac{du}{d\xi} d\xi.$$

Then T is a solvable extension of the minimal operator L_0:

$$T^{-1} f = \int_0^x f(\xi) \, d\xi - \frac{x}{b+1} \int_0^b f(\xi) \, d\xi,$$

which is not a restriction of $\tilde{L}$.

Example 2. Again let $V = (0, b)$, $L(D) = D_x$. The operator $T: \mathbb{H} \to \mathbb{H}$ defined as the closure in $\mathbb{H}$ of the operation $L(D)$, considered on the elements $u \in C^1$, and subject to the additional condition

$$u(0) = (q, L(D) u), \qquad q \in \mathbb{H},$$

(the parentheses denote the scalar product in $\mathbb{H}$) is, for every element $q \in \mathbb{H}$, a proper restriction of the operator L:

$$T^{-1} f = \int_0^x f(\xi) \, d\xi + (q, f).$$

But this operator will be an extension of L_0 if and only if $q = \text{const}$. (This example is a special case of one that will be discussed in subsection 1.2 of Chapter III).

It will be shown later that proper operators generated by an operation $L(D)$ exist under very broad assumptions about the operation. The chain of inclusions $L_0 \subset L \subset \tilde{L}$ and our examples address the point that a proper operator can be obtained either by removing part of the null boundary conditions imposed in the definition of L_0, or by restricting the operator L by imposing certain boundary conditions. A complete description of proper operators generated by ordinary differential operations $(n=1)$ will be given in the next chapter, precisely in terms of boundary conditions.

The classical boundary value problems of mathematical physics also furnish examples of proper operators. However, the general theorem on the existence of a proper operator, one of the proofs of which we are about to present, is established by transformation into an abstract theorem of functional analysis, and is, generally speaking, nonconstructive in the sense that it does not provide a method for describing the proper operators in terms of the boundary conditions. A different proof of the theorem, together with some of its realizations, is given in Chapter VII.

Lemma 1. *Let the closed linear operators* T_0 *and* $\tilde{T}$: $\mathbb{H}_1 \to \mathbb{H}_2$, $T_0 \subset \tilde{T}$, *have the property that* T_0 *has a bounded inverse* T_0^{-1}: $\mathbb{H}_2 \to \mathbb{H}_1$, *and the range* $\mathfrak{R}(\tilde{T})$ *of* $\tilde{T}$ *is the whole space* $\mathbb{H}_2$. *Then there is an operator* T: $\mathbb{H}_2 \to \mathbb{H}_1$, T_0 $\subset T \subset \tilde{T}$, *such that* T^{-1} *exists, is bounded, and is defined on the whole space* $\mathbb{H}_2$.

Proof. The conclusion is trivial if $\mathfrak{R}(T_0) = \mathbb{H}_2$: we have only to take $T = T_0$. Let $\mathfrak{R}(T_0)$ be a proper subspace of $\mathbb{H}_2$, and $\mathfrak{D}_0 \subset \mathbb{H}_1$ the largest linear manifold mapped by $\tilde{T}$ on $\mathfrak{R}(T_0)$. Evidently $\text{Ker } \tilde{T} \subset \mathfrak{D}_0$, and for $\tilde{T}$ considered on $\mathbb{H} \backslash \mathfrak{D}_0$ there is an inverse operator defined on $\mathbb{H}_2 \backslash \mathfrak{R}(T_0)$. Then the required operator T can be defined by setting $T = T_0$ on $\mathfrak{D}(T_0)$ and $T = \tilde{T}$ on $\mathbb{H}_1 \backslash \mathfrak{D}_0$. The boundedness of T^{-1} follows from Banach's theorem: subsection 1.3 of Chapter I.   □

We make some remarks. In what follows, we shall be interested only in the case $\mathbb{H}_1 = \mathbb{H}_2 = \mathbb{H}$; but in presenting the lemma it was convenient to operate with a pair of spaces.

We cannot define T simply by discarding the kernel of $\tilde{T}$: an operator constructed in this way will not, in general, be an extension of T_0.

An operator T with the required property of "regularity" is not, of course, uniquely determined. We shall see that even in the simplest cases there may be a continuum of different operators T.

Lemma 2. *For an operator* T: $\mathbb{H}_1 \to \mathbb{H}_2$ *with dense domain* $\mathfrak{D}(T) \subset \mathbb{H}_1$ *let there exist a bounded inverse* T^{-1}. *Then the equation*

$$T^* u = f$$

is solvable for every element $f \in \mathbb{H}_1$.

Proof. The conclusion of the lemma means that for every element $f \in \mathbb{H}_1$ there exists an element $u \in \mathbb{H}_2$ such that the equation

$$(T v, u)_2 = (v, f)$$

holds for every $v \in \mathfrak{D}(\mathbf{T})$. Introducing an element w: $\mathbf{T}v = w$, $v = \mathbf{T}^{-1}w$, we may write the chain of equations

$$(v, f)_1 = (\mathbf{T}^{-1}w, f)_1 = (w, u)_2 = (\mathbf{T}v, u)_2.$$

In passing from the second term to the third, we used the fact that the scalar product $(\mathbf{T}^{-1}w, f)$ defines, for a given element $f \in \mathbb{H}_1$, a bounded linear functional of w (that is, $|\mathbf{T}^{-1}w, f)_1| \leq \|\mathbf{T}^{-1}\| \|w\|_2 \|f\|_1$), which can be represented in the form of a scalar product $(w, u)_1$, where u is an element of $\mathbb{H}_2$ (Riesz's lemma, subsection 1.4 of Chapter I). Comparison of the first and last terms of the chain yields the conclusion of the lemma. $\square$

The next theorem follows from these lemmas and the considerations of the preceding subsection.

Theorem. *If the operation* $\mathbf{L}(D)$, *with sufficiently smooth coefficients, defined in a region* $V \subset \mathbb{R}^n$, *has the property that the operators* $\mathbf{L}_0$ *and* $\mathbf{L}_0^t$ *have bounded inverses and the equation* $\mathbf{L} = (\mathbf{L}_0^t)^*$ *is satisfied, then a proper operator* $\mathbf{L}$: $\mathbb{H}(V) \to \mathbb{H}(V)$ *generated by* $\mathbf{L}(D)$ *exists.*

Remark 1. A proof of this theorem for an arbitrary operation with constant coefficients, using the ideas of [44], was first obtained in [31]. There the existence of the bounded operator $\mathbf{L}_0^{-1}$ was deduced from the inequality

$$\|u\| \leq c \|\mathbf{L}_0 u\|, \tag{4}$$

which was established for $u \subset \mathfrak{D}(\mathbf{L}_0)$ by using the Fourier transform. In our later constructions (Chapters IV–VII) we shall typically use Fourier *series*, which in many respects are more natural for studying differential operations in a bounded domain, are more closely related to the classical methods of mathematical physics, and in particular permit a significant simplification in the proofs of inequalities like (4).

Remark 2. If we drop the assumption of the validity of the equation $\tilde{\mathbf{L}} = (\mathbf{L}_0^t)^*$, we can state a somewhat weaker conclusion: $\mathbf{L}(D)$ generates a solvable extension $\mathbf{L}_s$ of $\mathbf{L}_0$ such that

$$\mathbf{L}_0 \subset \mathbf{L}_s \subset (\mathbf{L}_0^t)^*.$$

The possibility of applying Lemmas 1 and 2 to the proof of the existence of proper operators is not, to be sure, exhausted by the cases in which the minimal operator plays the role of $\mathbf{T}_0$ in Lemma 1 and the maximal operator plays the role of $\tilde{\mathbf{T}}$. Later we shall meet examples of the use of these lemmas in other situations.

Here it is appropriate to repeat once again that the fundamental topic of the present monograph is the discovery of methods of effective description of proper operators generated by various classes of differential operations, and the study of the properties of these operators as they depend on properties of the initial operations and on the choice of the "boundary conditions" determining the operators.

§4. Weak and Strong Extensions of Differential Operations

4.0. Preliminary Remarks. As we have already noticed, the special place occupied by Hilbert space in the hierarchy of normed linear spaces can be attributed to its "selfadjointness": the possibility of identifying, in a natural way, the space of functionals with the original space (Riesz's lemma). Another aspect of the same selfadjointness is the existence in the algebra of operators on Hilbert space of an involution – the transition to an adjoint operator that belongs to the same algebra. The fundamental role of this operation is well known.

The concept of the adjoint of an operator inevitably also arises in the process of applying the methods of functional analysis to the study of specific objects, the operators generated by differential operators on the space of functions of integrable square.

In the preceding sections we have seen examples of the use of this concept. In the discussion of these examples there have already appeared various methods of defining the operators L and L^*: if the operator L: $\mathbb{H} \to \mathbb{H}$ generated by a differential operation is defined as the closure in $\mathbb{H}$ of this operation, considered initially on smooth functions, then the operator L^*: $\mathbb{H} \to \mathbb{H}$ is defined by an integral identity that yields, as one often says, the weak extension of the operator $L^t(D)$.

The definition of L^* by using an integral identity is inconvenient in many situations: the properties of an element $v \in \mathbb{H}$ which is a solution of the equation $L^* v = g$ often yield to investigation only with difficulty. In this connection there are special means for averaging (or regularizing), which can be used in a number of cases for establishing the equivalence of weak and strong (with respect to closure) definitions of operators generated by differential operations.

Sections 4–6 are devoted to the discussion of the corresponding circle of questions. In the present section we present the general definition of weak and strong extensions; in §5 we construct the averaging technique; and in §6 we establish a number of the equivalence theorems that were mentioned above.

4.1. Basic Definitions. In a domain $V \subset \mathbb{R}^n$ let there be given a differential operation $L(D)$ of the form (1), §2, with sufficiently smooth coefficients so that we can construct the operation $L^t(D)$. Let the original domain of $L(D)$ consist of the linear manifold $\mathscr{F}_\Gamma$ of smooth functions that in addition satisfy a system of homogeneous boundary conditions, which we write symbolically in the form

$$\Gamma u|_S = 0, \tag{Γ}$$

where S is the boundary of V.

A function $u \in \mathscr{F}_\Gamma$ that satisfies the equation

$$L(D) u = f, \tag{L}$$

will be called a *classical solution* of the $L - \Gamma$ problem.

An operator L_Γ: $\mathbb{H} \to \mathbb{H}$, defined as the closure in $\mathbb{H}$ of the operation $L(D)$, considered on $\mathscr{F}_\Gamma$, is called a *strong extension* of $L(D)$ under the conditions (Γ).

The terminology emphasizes, on the one hand, that a solution $u \in \mathbb{H}$ of the operator equation

$$L_\Gamma u = f$$

(a *strong solution* of the $L - \Gamma$ problem) is only a generalized solution, i.e. does not, in general, have all the derivatives that appear in $L(D)$, at least in the ordinary (classical) sense, and may fail to satisfy condition (Γ). On the other hand, the extension (or solution) is called strong since one often uses weaker definitions of extensions of classes of operations $L(D)$ or of solutions of the $L - \Gamma$ problem.

Such weak extensions $\tilde{L}_\Gamma \supset L$ are most often defined in the following way. Let

$$t \gamma v|_S = 0 \tag{$t\gamma$}$$

be a system of homogeneous boundary conditions such that smooth functions u and v that satisfy conditions (Γ) and $(t\gamma)$, respectively, satisfy the equation

$$(L(D) u, v) = (u, L^t(D) v). \tag{1}$$

Let $L^t_{t\gamma}$: $\mathbb{H} \to \mathbb{H}$ be the operator generated by the operation $L^t(D)$ and the conditions $(t\gamma)$ in the same way as L_Γ is generated by $L(D)$ and the conditions (Γ) (i.e. the closure of the operation $L^t(D)$ originally defined on the linear manifold $\mathscr{F}_{t\gamma}$ of smooth functions). Then equation (1) immediately implies the inclusion

$$L_\Gamma \subset (L^t_{t\gamma})^* \tag{2}$$

(cf. §3), i.e. the operator $(L^t_{t\gamma})^*$ can be taken to be the operator $\tilde{L}_\Gamma \supset L_\Gamma$ that was discussed above.

In the majority of most interesting cases it is possible to determine a "minimal" system γ of homogeneous boundary conditions such that, for smooth functions u satisfying the equation,

$$\gamma u|_S = 0 \tag{γ}$$

is a *necessary* condition for (1) to hold for every smooth v that satisfies $(t\,\gamma)$. Then the operator $(\mathbf{L}^t_{t\,\gamma})^*$ is called a *weak extension* of $\mathbf{L}(D)$ *under conditions* (Γ).

The existence of the embedding (2) means that when conditions (Γ) are satisfied, condition (γ) follows $(\mathscr{F}_\Gamma \subset \mathscr{F}_\gamma)$.

In a number of important cases it is possible to establish the equation

$$\mathbf{L}_\Gamma = (\mathbf{L}^t_{t\,\gamma})^*. \tag{3}$$

Then it is natural to expect that conditions (Γ) and (γ) are the same. But this is not necessarily the case. Conditions (Γ) may contain redundant requirements that have no real influence on the properties of elements of $\mathfrak{D}(L_\Gamma)$ ("are not preserved in the closure").

If, for example, $V=(0, b)$ and $L(D)=D_x$, the manifold $\mathscr{F}_{\Gamma_1}$ consists of the elements $u \in C^\infty(\bar{V})$ that satisfy the additional condition

$$u|_0 = u'|_0 = \ldots = u^{(k)}|_0 = 0, \qquad k \geq 1,$$

and the manifold $\mathscr{F}_{\Gamma_2}$ consists of the elements $w \in C^1(\bar{V})$ that satisfy the condition $w|_0 = 0$, then, as is well known (and will follow from the subsequent discussion) the operators $\mathbf{L}_{\Gamma_1}$ and $\mathbf{L}_{\Gamma_2}$ are the same, and if conditions $(t\,\gamma)$ have the form $v|_{x=b} = 0$, then both operators will satisfy (3).

The equations (2) and (3) of §3 above are special cases of (3). Here if (Γ) are the conditions defining the minimal operator, and $(t\,\gamma)$ are the absence of any conditions, then, as in the preceding example, the presence of (3) does not mean that (Γ) and (γ) coincide.

If together with conditions (Γ), (γ) and $(t\,\gamma)$ we can introduce the conditions $(t\,\Gamma)$ (related to (γ) in the same way that $(t\,\gamma)$ are related to (Γ)), and also

$$\Gamma = \gamma, \qquad t\,\gamma = t\,\Gamma, \tag{4}$$

then conditions (Γ) and $(t\,\Gamma)$ are said to be *adjoint*.

The example in which

$$u|_0 = 0, \qquad (\Gamma) \qquad\qquad u|_{x=b} = 0, \qquad (t\,\Gamma)$$

$$v|_{x=b} = v'|_{x=b} = 0, \qquad (t\,\gamma) \qquad\qquad v|_0 = 0, \qquad (\gamma)$$

shows that the validity of the first of equations (4) does not guarantee the validity of the second.

A reader who is familiar with the operational formalism will, of course, have noticed that in equation (3) the operator $(\mathbf{L}^t_{t\,\gamma})^*$ plays the role of the "second adjoint" with respect to $\mathbf{L}_\Gamma$, i.e. we are dealing with a variation on the theme of the classical equation $\mathbf{T} = \mathbf{T}^{**}$.

§ 5. Averaging Operators

5.0. Preliminary Remarks. Averaging operators provide a technique for replacing an element of some function space by its regularization, i.e. an element of the same space that has more smoothness (or regularity) and at the same time is close to it in the corresponding norm.

An averaging operator is usually defined as the convolution of a given element with a function defined by a smooth approximation of the Dirac δ-function (which plays the role of a unity in the algebra with convolution as multiplication). Depending on the nature of the problem under investigation, the regularization to be used may involve some special requirement, which determines the choice of the averaging operator.

In §§ 5 and 6 we shall suppose, as is usual in considering these problems, that $\mathbb{H}$ is real Hilbert space. The transition, if necessary, to the complex case can be made automatically.

5.1. Averaging on the Line. Let $\omega(\xi) \in C^\infty$ be an even nonnegative function such that $\omega(\xi) = 0$ for $|\xi| \geq 1$ and $\int \omega(\xi) \, d\xi = 1$ (if the limits of integration are not shown, the integration is extended over the whole real line). The simplest example of a function $\omega(\xi)$ with the necessary properties, known as an *averaging kernel*, is $\omega(\xi) = \alpha(\xi) / \int \alpha(\xi) \, d\xi$, where $\alpha(\xi) = \exp\left(\dfrac{1}{\xi^2 - 1}\right)$ for $|\xi| < 1$ and $\alpha(\xi) = 0$ for $|\xi| \geq 1$.

Let $\omega_\varepsilon(x, x')$ denote the function $\varepsilon^{-1} \omega\left(\dfrac{x - x'}{\varepsilon}\right)$. For arbitrary $\varepsilon > 0$ we have

$$\int \omega_\varepsilon(x, x') \, dx = \int \omega_\varepsilon(x, x') \, dx' = 1$$

and $\omega_\varepsilon(x, x') = 0$ for $|x - x'| \geq \varepsilon$.

Now let $V = (0, b)$ be a finite real interval and $u \in \mathbb{H} \equiv \mathbb{H}(V)$. For every $\varepsilon > 0$ we define an operator $J_\varepsilon \colon \mathbb{H} \to \mathbb{H}$ by the equation

$$J_\varepsilon u(x) = \int_V \omega_\varepsilon(x, x') \, u(x') \, dx'.$$

The operator J_ε (and modifications of it that we shall meet later) is called an *averaging operator*. Let us establish some fundamental properties of such operators.

J-1. *For every element $u \in H$, we have $J_\varepsilon u \in C^\infty$, and*

$$D_x^m J_\varepsilon u = \int_V [D_x^m \omega_\varepsilon(x, x')] \, u(x') \, dx'. \tag{1}$$

Proof. We can prove this by induction. If (1) holds for $m - 1$, it is enough to observe that, for real $\delta > 0$,

$$\delta^{-1}[D^{m-1}J_\varepsilon u(x+\delta)-D^{m-1}J_\varepsilon u(x)]$$
$$=\delta^{-1}\int_V [D_x^{m-1}\omega_\varepsilon(x+\delta,x')-D_x^{m-1}\omega_\varepsilon(x,x')]\,u(x')\,dx'$$
$$=\int_V [D_x^m \omega_\varepsilon(x+q,x')]\,u(x')\,dx',$$

and then estimate the difference between the last term of this chain of equations and the right-hand side of (1) as $\delta\to 0$, using the smoothness of $\omega(\xi)$. $\square$

J-2. $\|J_\varepsilon\|\le 1$ for every $\varepsilon>0$.

We prove J-2 by establishing a somewhat more general result.

Lemma 1. *Let* $\mathbb{K}$: $\mathbb{H}\to\mathbb{H}$ *be the integral operator*

$$\mathbf{K}u=\int_V K(x,x')\,u(x')\,dx',$$

with a continuous kernel that satisfies the conditions

$$\int_V |K(x,x')|\,dx\le M, \qquad \int_V |K(x,x')|\,dx'\le M.$$

Then $\|\mathbf{K}\|\le M$.

Proof. It is sufficient to observe that

$$\|\mathbf{K}u\|^2=\int_V|\int_V K(x,x')\,u(x')\,dx'|^2\,dx$$
$$\le\int_V(\int_V |K(x,x')|\,dx'\int_V |K(x,x')|\,|u(x')|^2\,dx')\,dx\le M\int_V |u(x')|^2\,dx'.$$

Here we wrote $|K|=|K|^{\frac{1}{2}}|K|^{\frac{1}{2}}$, used the Cauchy-Bunyakovsky inequality, and changed the order of integration.

Now J-2 follows immediately from the lemma and the properties of $\omega(\xi)$.

Remark. Both Lemma 1 and its proof evidently remain valid when V is any bounded domain in $\mathbb{R}^n$. We shall use this fact in what follows. Generally speaking, the extension of these properties to the case $n>1$ presents no difficulties.

J-3. *When* $\varepsilon\to 0$ *the norm of the difference* $J_\varepsilon u-u$ *tends to zero.*

Proof. To estimate $\|J_\varepsilon u-u\|^2$ (norm in $\mathbb{H}$) it is evidently enough to estimate the integral

$$I(u)=\int_V|\int_V \omega_\varepsilon(x,x')\,[u(x')-u(x)]\,dx'|^2\,dx$$
$$\le\int_V\{\int_V |\omega_\varepsilon(x,x')|^2\,dx'\int_{|x-x'|\le\varepsilon} |u(x')-u(x)|^2\,dx'\}\,dx.$$

In order to attach a precise meaning to the second integral in the braces for every $x \in V$, we suppose that $u(x) \equiv 0$ for $x \notin V$. Evidently $\int_V |\omega_\varepsilon(x, x')|^2 \, dx$ $\leq c \varepsilon^{-1}$ uniformly in x, and the constant c depends only on the form of $\omega(\xi)$. If we now set $x' = x + \tau$, $|\tau| \leq \varepsilon$, $dx' = d\tau$, we will have

$$I(u) \leq c \varepsilon^{-1} \int_V \left(\int_{|\tau| \leq \varepsilon} |u(x + \tau) - u(x)|^2 \, d\tau \right) dx$$

$$= c \varepsilon^{-1} \int_{|\tau| \leq \varepsilon} \left(\int_V |u(x + \tau) - u(x)|^2 \, dx \right) d\tau \leq c' \mathscr{B}_\varepsilon u,$$

where $\mathscr{B}_\varepsilon u \to 0$ as $\varepsilon \to 0$ (continuity in the mean: subsection 1.1). $\quad\square$

Remark. It is easy to see that for any given $\varepsilon > 0$ the operator J_ε carries a bounded set $\mathscr{M} \subset \mathbb{H}$ to a set of elements $J_\varepsilon \mathscr{M}$ that are uniformly bounded and equicontinuous. Hence it follows that $J_\varepsilon \mathscr{M}$ is a compact subset of C (Arzelà's theorem), and a fortiori in $\mathbb{H}$. Therefore for a given $\varepsilon > 0$ the operator $J_\varepsilon \colon \mathbb{H} \to \mathbb{H}$ (like every integral operator with a smooth kernel) is completely continuous.

J-4. *If $a(x) \in C(V)$ then*

$$\| a J_\varepsilon u - J_\varepsilon (a u) \| \to 0 \qquad as \ \varepsilon \to 0$$

for every $u \in \mathbb{H}$.

In fact, it is sufficient to observe that

$$\| a J_\varepsilon u - J_\varepsilon (a u) \| \leq \| a(J_\varepsilon u - u) \| + \| a u - J_\varepsilon (a u) \|. \quad \square$$

Property J-4 shows that "in the limit" the averaging operator commutes with the operation of multiplication by a sufficiently regular function.

J-5. *For all u and $v \in \mathbb{H}$ we have the equation*

$$(J_\varepsilon u, v) = (u, J_\varepsilon v). \tag{2}$$

Equation (2) follows at once from the possibility of changing the order of integration.

Later on, we shall need averaging operators with the property that the smooth functions constructed by using them also satisfy some homogeneous boundary conditions. In the one-dimensional case the simplest such operators are, for example,

$$J_\varepsilon^- u = \int_V \omega_\varepsilon \left(\frac{x - x' - 2\varepsilon}{\varepsilon} \right) u(x') \, dx',$$

$$J_\varepsilon^+ u = \int_V \omega_\varepsilon \left(\frac{x - x' + 2\varepsilon}{\varepsilon} \right) u(x') \, dx'. \tag{3}$$

Properties J-1, 2, 3, 4 are established for these operators exactly as for J_ε. Property J-5 has a somewhat different appearance.

J-5a. *For all u and $v \in \mathbb{H}$ we have the equation*

$$(J_\varepsilon^- u, v) = (u, J_\varepsilon^+ v).$$

But in addition we have still another property.

J-6a. *For every element $u \in \mathbb{H}(V)$ we have the equations*

$$J_\varepsilon^- u|_{x=0} = 0, \qquad J_\varepsilon^+ u|_{x=b} = 0.$$

This follows immediately from (3): for $x = 0$ (or $x = b$) the kernel of J_ε^- (or J_ε^+) is identically zero for every $x' \in V = (0, b)$.

It is also easy to construct an operator that satisfies homogeneous conditions at both ends of the interval:

$$\dot{J}_\varepsilon u = \int_V \omega_\varepsilon \left(\frac{x - (1 - 4b^{-1}\varepsilon) x' - 2\varepsilon}{\varepsilon} \right) u(x') \, dx'.$$

Here property J-1 is unchanged. To prove J-2 and J-3 we have to replace the unit Jacobian by the Jacobian $1 - 4b^{-1}\varepsilon$; this makes no difference. In property J-2 we have to replace the unit constant in the inequality by a number $E_\varepsilon \to 1$ as $\varepsilon \to 0$. It is necessary only to change the forms of J-5 and J-6.

J-5b. *For every u and $v \in \mathbb{H}$ we have the equation*

$$(\dot{J}_\varepsilon u, v) = (u, \tilde{J}_\varepsilon v),$$

where

$$\tilde{J}_\varepsilon v = \int_V \omega_\varepsilon \left(\frac{(1 - 4b^{-1}\varepsilon) x - x' + 2\varepsilon}{\varepsilon} \right) v(x') \, dx'.$$

In addition, we have the property

J-6b. *For every element $u \in \mathbb{H}(V)$ we have the equation*

$$\dot{J}_\varepsilon u|_{x=0} = \dot{J}_\varepsilon u|_{x=b} = 0.$$

The operator $\tilde{J}_\varepsilon$ "spreads out" the averaged element (as does the original operator J_ε), and $\tilde{J}_\varepsilon u$ does not in general satisfy homogeneous boundary conditions at either end of the interval $[0, b]$.

5.2. Averaging in a Multidimensional Domain. Turning to the case when V is a bounded domain in $\mathbb{R}^n$, we define the operator J_ε by the equation

$$J_\varepsilon u(x) = \int_V \omega_\varepsilon(x_1, x_1') \ldots \omega_\varepsilon(x_n, x_n') u(x') \, dx', \tag{4}$$

where $x = (x_1, \ldots, x_n)$ and x', dx' are defined similarly.

The statements of properties J-1 to J-5 remain as before (except that naturally D^m is replaced by the partial derivative D^α, where α is a multi-index: §2). The proofs also remain valid (with the evident change of the inequality $\int_V |\omega_\varepsilon|^2 \, dx' \le c\varepsilon^{-1}$ to $\le c\varepsilon^{-n}$ in the proof of property J-3, etc.).

We shall not use operators of the type of J^- and J^+ in the multidimensional case, and give details only for operators of the type of $\overset{\ast}{J}_\varepsilon$ and $\tilde{J}_\varepsilon$.

Let the domain V have the property that there is a family $\{\varphi^\varepsilon(x)\}$, $0 \le \varepsilon \le \varepsilon_0$, of diffeomorphisms $\varphi^\varepsilon \colon V \to V^\varepsilon \subset V$ of V to a subset, depending smoothly on ε and such that

V-1. *The mapping* $\varphi^0 = 1$ *is the identity; the Jacobians* $j(\varphi^\varepsilon)$ *and* $j^{-1}(\varphi^\varepsilon)$ *tend, uniformly in* $x \in V$, *to unity as* $\varepsilon \to 0$.

V-2. *If* $x \in S = \partial V$, *then for each* $x' \in V^\varepsilon$ *the Euclidean distance* $d(x, x')$ *satisfies the inequality* $d(x, x') \ge k\varepsilon$, *where* $k \ge 1$ *is a fixed number.*

A domain V that satisfies conditions V-1 and V-2 is usually called *normal*.

The conditions on the boundary of V at the beginning of this chapter (subsection 1.1) are automatically sufficient for normality.

Remark. A somewhat less restrictive condition that guarantees normality can be stated as follows. A domain V_k is *star-shaped* with respect to a ball $S_k \subset V_k$ (the radius of the ball is positive) if, for every point $x \in \partial V_k$ the cone based on S_k and with vertex x lies entirely in $\bar{V}_k$. A domain V which is the union of a finite number of domains V_k, each of which is star-shaped with respect to a ball S_k, is normal.

In the one-dimensional case considered above $(V = (0, b))$ we used mappings φ^ε of the form

$$\varphi^\varepsilon(x) = (1 - 4b^{-1}\varepsilon) x + 2\varepsilon.$$

If we take $\varphi^\varepsilon = (\varphi_1^\varepsilon, \ldots, \varphi_n^\varepsilon)$, then the operator $\overset{\ast}{J}_\varepsilon$ can be defined by the equation

$$\overset{\ast}{J}_\varepsilon u(x) = \int_V \omega_\varepsilon \left(\frac{x_1 - \varphi_1^\varepsilon(x')}{\varepsilon} \right) \ldots \omega_\varepsilon \left(\frac{x_n - \varphi_n^\varepsilon(x')}{\varepsilon} \right) u(x') \, dx'.$$

The operator $\overset{\ast}{J}_\varepsilon$ is an averaging operator that is fully analogous to the corresponding operator in the one-dimensional case. We state property J-5b (supposing, naturally, that $0 < \varepsilon \le \varepsilon_0$).

J-5b. *For all* $u, v \in \mathbb{H}(V)$ *we have the equation*

$$(\overset{\ast}{J}_\varepsilon u, v) = (u, \tilde{J}_\varepsilon v),$$

where

$$\tilde{J}_\varepsilon v = \int_V \omega_\varepsilon \left(\frac{\varphi_1^\varepsilon(x) - x_1'}{\varepsilon} \right) \ldots \omega_\varepsilon \left(\frac{\varphi_n^\varepsilon(x) - x_n'}{\varepsilon} \right) v(x') \, dx'.$$

We again find that $\tilde{J}_\varepsilon$ turns out to be a "spreading out" averaging operator, and that like $\dot{J}_\varepsilon$ it possesses property J-6b.

J-6b. *For every $u \in \mathrm{IH}(V)$ we have the equation*

$$\dot{J}_\varepsilon u|_{\partial V} = 0.$$

5.3. Averaging and the Differentiation Operation. As we observed above, averaging is the fundamental tool that lets us establish the equivalence of weak and strong extensions of differential operations. We now present the general method for using averaging for such purposes.

Let D^α be a differential monomial of the form $D_1^{\alpha_1} \ldots D_n^{\alpha_n}$. An element $u \in \mathrm{IH}(V)$ belongs to $\mathfrak{D}(D_{\mathrm{wk}}^\alpha)$ (the domain of D^α, taken in the weak sense) if there is an element $f \in \mathrm{IH}(V)$ such that for every "admissible" function $\varphi(x)$ we have the equation

$$\int_V u \, D^\alpha \varphi \, dV = (-1)^{|\alpha|} \int_V f \varphi \, dV. \tag{5}$$

The definition of admissibility, in addition to the necessary requirement of sufficient smoothness, can contain in each separate case one or another supplementary condition on the boundary. (We shall discuss this aspect of the question in more detail in the next section.)

Now let an averaging operator J_ε (not necessarily the same as the operator so denoted in (4)) assign to every element $v \in \mathrm{IH}(V)$ an admissible function $\varphi = J_\varepsilon v$. Then if we set $\varphi = J_\varepsilon v$ in (5), we obtain the equation

$$(u, D^\alpha J_\varepsilon v) = (-1)^{|\alpha|} (f, J_\varepsilon v) \tag{6}$$

(the parentheses denote the scalar product in IH).

The term $D^\alpha J_\varepsilon v$ on the left of (6) can be taken to be the result of operating on v with the integral operator $D^\alpha J_\varepsilon$ with kernel $D_x^\alpha \omega_\varepsilon(x, x')$:

$$D^\alpha J_\varepsilon v = \int_V [D_x^\alpha \omega_\varepsilon(x, x')] \, v(x') \, dx' = [D^\alpha J_\varepsilon] \, v.$$

As follows from the definition of an averaging operator, for every $\varepsilon > 0$ this integral operator has an adjoint $[D^\alpha J_\varepsilon]^*$, and

$$(u, [D^\alpha J_\varepsilon] \, v) = ([D^\alpha J_\varepsilon]^* \, u, v) = (-1)^{|\alpha|} (D^\alpha [J_\varepsilon^t u], v), \tag{7}$$

where the last term of the equation is obtained by applying the evenness of $\omega(\xi)$ and replacing $D^\alpha \omega_\varepsilon(x, x')$ by $(-1)^{|\alpha|} D_{x'}^\alpha \omega_\varepsilon(x, x')$ (possibly with the introduction of an auxiliary factor $E_\varepsilon \to 1$ as $\varepsilon \to 0$, included in the definition of J_ε^t; cf. the preceding subsections).

The operator J_ε^t is also an averaging operator which maps IH on the linear manifold $\mathcal{M} \subset \mathrm{IH}$ of smooth functions that may also satisfy some auxiliary condition on the boundary. If we now set

$$\varepsilon_k = 2^{-k}, \quad k = 1, 2, \ldots; \qquad u_k = J^l_{\varepsilon_k} u \to u \quad \text{as } \varepsilon_k \to 0,$$

then by (6), (7), and the arbitrary nature of $v \in \mathbb{H}$,

$$D^\alpha u_k = J^*_{\varepsilon_k} f \to f \quad \text{as } k \to \infty,$$

i.e. $u \in \mathfrak{D}(D^\alpha_{wk})$ implies $u \in \mathfrak{D}(D^\alpha_{\mathcal{M}})$, the domain of the operator $D^\alpha_{\mathcal{M}}$ previously defined as the closure in $\mathbb{H}$ of the operation D^α, initially defined on the linear manifold $\mathcal{M}$.

Remark. Considering the smooth function φ in (5) to be admissible if it is finitely supported in V (equal to zero outside a compact subset $V' \subset V$), we arrive at the definition of the generalized derivative $D^{\alpha_1}_1 \ldots D^{\alpha_n}_n$ adopted in [20]. Supposing that V is normal, and taking J_ε to be the operator $\mathring{J}_\varepsilon$ (subsection 5.2), we obtain the inclusion $\mathfrak{D}(D^\alpha_{wk}) \subset \mathfrak{D}(\tilde{D}^\alpha)$, where $\tilde{D}^\alpha$ is the corresponding maximal operator.

5.4. The Friedrichs Lemma. This lemma will be used in parts of the next section that are somewhat outside the mainstream of the exposition, but have substantial interest in principle.

The conclusion of the lemma is directly related to property J-4 of subsection 5.1, and says that the permutability (in the limit) of averaging with multiplication by a smooth function can also take place "inside the sign of differentiation." As in subsection 5.1, we confine ourselves to the case $n = 1$, $V = (0, b)$. The transfer of the construction to the case of arbitrary n presents no difficulty. First we state an auxiliary result.

Proposition 1. *Let* $\mathbf{K}_\varepsilon$: $\mathbb{H} \to \mathbb{H}$ *be a family of integral operators of the form* $\mathbf{K}_\varepsilon u = \int_V K_\varepsilon(x, x') u(x') dx'$, *and also*

K-1. *There is a constant M, independent of ε, such that* $\|\mathbf{K}_\varepsilon\| \leq M$.

K-2. *For some number κ, arbitrary but sufficiently small τ and $x \in V_\tau = (\tau, b - \tau)$, there is an $\varepsilon' = \varepsilon'(\tau)$ such that*

$$\int_V K_\varepsilon(x, x') dx' = \kappa \quad \text{for all } \varepsilon < \varepsilon'(\tau). \tag{8}$$

K-3. $K_\varepsilon(x, x') \equiv 0$ *for* $|x - x'| > k \varepsilon$, $k = \text{const}$. *Then* $\|\mathbf{K}_\varepsilon u - \kappa u\| \to 0$ *as* $\varepsilon \to 0$.

The conclusion is a generalization of property J-3 of 5.1 (where $\kappa = 1$), and is established by the same reasoning (in addition, we need only use the approach to zero, as $\tau \to 0$, of the integral over $V \backslash V_\tau$, where (8) need not hold). $\square$

Lemma (Friedrichs). *Let* $u \in \mathbb{H}(V)$, $a \in C^1(V)$, *and let* J_ε *be a standard averaging operator. Then*

$$\|D(a J_\varepsilon u) - D J_\varepsilon(a u)\| \to 0 \quad \text{as } \varepsilon \to 0.$$

Proof. First we observe that

$$DJ_\varepsilon(a\,u) = D \int_V \omega_\varepsilon(x, x')\, a(x')\, u(x')\, dx = - \int_V (D'\,\omega_\varepsilon)\, a(x')\, u(x')\, dx'$$

$$= - \int_V \{D'[\omega_\varepsilon a(x')]\}\, u(x')\, dx' + \int_V \omega_\varepsilon \cdot [D' a(x')]\, u(x')\, dx', \qquad (9)$$

where D' is differention with respect to x'. We also have

$$D(a\,J_\varepsilon\,u) = D a \cdot \int_V \omega_\varepsilon \cdot u(x')\, dx' - \int_V [D'\,\omega_\varepsilon]\, a(x)\, u(x')\, dx'. \qquad (10)$$

The difference between the last term of (9) and the first term on the right of (10) evidently tends to zero (interchange of averaging and multiplication by a continuous function). We can write the difference of the remaining terms in the form

$$\int_V D'\{\omega_\varepsilon(x, x')\, [a(x) - a(x')]\}\, u(x')\, dx'$$

and apply the preceding proposition, taking $K_\varepsilon = D'\{\omega_\varepsilon[a - a']\}$. The corresponding family of operators is uniformly bounded (we use the fact that $|a - a'| \le C\varepsilon$ for $|x - x'| \le \varepsilon$), satisfies K-2 with $\kappa = 0$, and satisfies K-3 locally. $\square$

This lemma has many variants. It is useful to notice that it is enough to require that $a(x)$ is piecewise differentiable and piecewise Lipschitzian.

It follows immediately from the Friedrichs lemma that when $u \in \mathfrak{D}(D_{wk})$ and $a(x) \in C^1$ (and consequently $a\,u \in \mathfrak{D}(D_{wk})$), we have

$$\|D(a\,J_\varepsilon\,u) - J_\varepsilon D_{wk}(a\,u)\| \to 0 \qquad \text{as } \varepsilon \to 0.$$

§6. The Identity of Weak and Strong Extensions of Differential Operations

6.1. The Case of Mainly Constant Coefficients. For operations $L(D)$ of the form (1), §2, whose coefficients a_α, $|\alpha| \ge 2$, are constants, the question of the equivalence of weak and strong extensions involving averaging operators which are well adapted to the boundary value problems under consideration can be answered quite simply because of the interchangeability of averaging and multiplication by a constant, and the approximate interchangeability of averaging with multiplication by a smooth function, in cases covered by property J-4, §5, or by the Friedrichs lemma. What is meant by "adapted" for boundary conditions will become clear in the course of the discussion.

We suppose that there are given a domain $V \subset \mathbb{R}^n$ and an operation $\mathbf{L}(D)$ whose coefficients satisfy the hypotheses stated above. In the notation of §4 let γ and $t\gamma$ be adjoint systems of boundary conditions. Let the averaging operator $J_\varepsilon^{t\gamma}$ have the property that for every $v \in \mathbb{H}(V)$ and every $\varepsilon > 0$, the function $J_\varepsilon^{t\gamma} v$ is smooth and satisfies the boundary conditions $t\gamma$. Then if $u \in \mathscr{D}(\mathbf{L}_\gamma^{wk})$ (the domain of the weak extension of $\mathbf{L}(D)$ under conditions γ: $\mathbf{L}_\gamma^{wk} \equiv (\mathbf{L}_{t\gamma}^t)^*$) and $\mathbf{L}_\gamma^{wk} u = f$, then, for every $v \in \mathbb{H}$,

$$(u, \mathbf{L}^t(D) J_\varepsilon^{t\gamma} v) = (f, J_\varepsilon^{t\gamma} v). \tag{1}$$

Next we apply the procedure described in subsection 3 of §5. Let the integral operator $\mathbf{L}^t(D) J_\varepsilon^{t\gamma}$ have the property that the adjoint operator satisfies the equation

$$[\mathbf{L}^t(D) J_\varepsilon^{t\gamma}]^* u = \mathbf{L}(D) J_\varepsilon^\gamma u + \eta_\varepsilon(u),$$

and let the following conditions be satisfied:

1) J_ε^γ is an averaging operator with the property that for every $\varepsilon > 0$ the smooth function $J_\varepsilon^\gamma u$ satisfies conditions γ for every $u \in \mathbb{H}$;

2) $\|\eta_\varepsilon(u)\| \to 0$ as $\varepsilon \to 0$, for every $u \in \mathbb{H}$.

Then (1) immediately reduces to the equation

$$\mathbf{L}(D) J_\varepsilon^\gamma u = J_\varepsilon^\gamma f - \eta_\varepsilon(u),$$

where the right-hand side approaches f as $\varepsilon \to 0$.

Defining the sequence $\{u_k\}$ as in subsection 3 of §5, we obtain the inclusion

$$\mathscr{D}(\mathbf{L}_\gamma^{wk}) \subset \mathscr{D}(\mathbf{L}_\gamma).$$

The converse inclusion is, as usual, evident.

If we now take $\mathbf{L}_\gamma$ to be the maximal operator generated by the operation $\mathbf{L}(D)$. and $J_\varepsilon^{t\gamma}$ to be the operator $\dot{J}_\varepsilon$ defined in subsection 2 of §5, we obtain the following theorem.

Theorem. *For an operation $\mathbf{L}(D)$ whose coefficients a_α are constant for $|\alpha| \geq 2$, the equation*

$$\tilde{\mathbf{L}} = (\mathbf{L}_0^t)^*$$

holds in a normal domain V.

6.2. The Case of Variable Coefficients. For a general differential operation $\mathbf{L}(D)$ of the form (1), §2, with variable coefficients a_α, $|\alpha| > 1$, the question of whether $\mathbf{L}_\gamma$ and $\mathbf{L}_\gamma^{wk}$ are the same is considerably more complex. The fundamental reason for the complication is the inadmissability, in general, of interchanging multiplication (by a smooth function) and averaging, under the differentiation sign of order greater than 1, i.e. the absence in this case of an analog of the Friedrichs lemma.

Let us look at the simplest example that illustrates the nature of the difficulty that arises. Take the averaging operator

$$\mathbf{K}_h u(x) = \frac{1}{2h} \int_{x-h}^{x+h} u(\xi)\, d\xi,$$

which makes an absolutely continuous function (the Steklov function) correspond to an element $u \in \mathbb{H}$. Then if $a = a(x) \in C^1$, we have

$$D(a\,\mathbf{K}_h u) = a'(x)\,\mathbf{K}_h u + \frac{a(x)}{2h}\,[u(x+h) - u(x-h)].$$

On the other hand, $D\mathbf{K}_h(a u)$ can be represented in the form

$$D\mathbf{K}_h(a u) = \frac{a(x+h) - a(x-h)}{2h}\,u(x+h) + \frac{a(x-h)}{2h}\,[u(x+h) - u(x-h)]$$

and it is evident that as $h \to 0$,

$$\|D(a\,\mathbf{K}_h u) - D\mathbf{K}_h(a u)\| \to 0$$

for every element $u \in \mathbb{H}$. This is the mechanism of the Friedrichs lemma.

If, however, supposing $a = a(x) \in C^2$, we try to compare

$$D^2(a\,\mathbf{K}_h^2 u) = D^2 a \cdot \mathbf{K}_h^2 u + 2Da \cdot D\mathbf{K}_h^2 u + a\,D^2 \mathbf{K}_h^2 u$$

and $D^2 \mathbf{K}_h^2(a u)$, representing this expression in the form

$$\begin{aligned}
4h^2 D^2 \mathbf{K}_h(a u) =\ & [a(x+2h) - 2a(x) + a(x-2h)]\,u(x+2h) \\
& + [a(x) - a(x-2h)]\,[u(x+2h) - u(x)] \\
& + a(x-2h)\,[u(x+2h) - 2u(x) + u(x-2h)],
\end{aligned}$$

then in order to estimate the norm

$$\|D^2(a\,\mathbf{K}_h^2 u) - D^2 \mathbf{K}_h^2(a u)\| \tag{2}$$

it is not enough to apply the factor $(2h)^{-2}$ (coming from $\mathbf{K}_h^2$) to the second differences of $a(x)$, and the norm (2) will approach zero as $h \to 0$ only under additional hypotheses on $u(x)$.

The preceding discussion does not, of course, contradict the fact that, as will be shown below, in the case of ordinary differential operations the weak and strong extensions are always the same.

Let us consider the nature of the additional hypotheses under which we can establish an analog of the Friedrichs lemma. Let, for example, $L(D)$ be an operation of order m and let it follow from $u \in \mathfrak{D}(\mathbf{L}_\gamma^{wk})$ that there is an element $f^\beta \in \mathbb{H}$ such that for each $v \in \mathbb{H}$ the inequality

$$(u, D^\beta J_\varepsilon^{t\gamma} v) = (-1)^{|\beta|} (f^\beta, J_\varepsilon^{t\gamma} v) \tag{3}$$

is satisfied for every multi-index β, $|\beta| \leq m-1$. In other words, from the fact that $u \in \mathfrak{D}(\mathbf{L}_\gamma^{wk})$ there follows the existence of all weak derivatives of u of order up to $m-1$ (the definition of the weak derivatives takes account of the boundary conditions γ). The averaging operator $J_\varepsilon^{t\gamma} v$ that we use ordinarily occurs in such a way that (3) also remains valid both if we replace $J_\varepsilon^{t\gamma} v$ by $a(x) J_\varepsilon^{t\gamma} v$ (where $a(x)$ is a smooth function) and even by $D_k[a(x) J_\varepsilon^{t\gamma} v]$ (the homogeneous boundary conditions for $J_\varepsilon^{t\gamma} v$ are satisfied identically, so that multiplication by a sufficiently smooth function and even differentiation do not interfere with their being satisfied).

We suppose that the averaging operator has the indicated property, and in the equation

$$(u, \mathbf{L}^t J_\varepsilon^{t\gamma} v) = (f, J_\varepsilon^{t\gamma} v), \qquad \varepsilon > 0 \tag{4}$$

(valid for every element $v \in \mathbb{H}$ and implying that $\mathbf{L}_\gamma^{wk} u = f$) we consider a term of the left-hand side, one containing, for example, the derivative D^α of highest order, $|\alpha| = m$. Let $D^\alpha = D^\beta D_k$, $|\beta| = m-1$. Then

$$(-1)^{|\alpha|} (u, D^\alpha[a_\alpha J_\varepsilon^{t\gamma} v]) = (-1)^{|\alpha|} (u, D^\beta[D_k(a_\alpha J_\varepsilon^{t\gamma} v)])$$
$$= -(f^\beta, D_k[a_\alpha J_\varepsilon^{t\gamma} v]) = (a_\alpha D_k J_\varepsilon^\gamma f^\beta + \eta_\varepsilon, v) = (a_\alpha D^\alpha J_\varepsilon^\gamma u + \eta_\varepsilon, v), \tag{5}$$

where $\|\eta_\varepsilon\| \to 0$ as $\varepsilon \to 0$. To obtain the chain (5) of equations we used the Friedrichs lemma and the equation $J_\varepsilon^\gamma f^\beta = D^\beta J_\varepsilon^\gamma u$, which follows from (3).

The chain (5) of equations, which plays the role of the Friedrichs lemma, allows us to show directly that under our hypotheses the weak and strong definitions of the operator $\mathbf{L}_\gamma : \mathbb{H} \to \mathbb{H}$ are equivalent. In fact, if we transform all the terms of the left side of (4) in the same way as in (5), we obtain the equation

$$(\mathbf{L} J_\varepsilon^\gamma u + \tilde{\eta}_\varepsilon, v) = (J_\varepsilon^\gamma f, v),$$

which, as in the case of constant coefficients, yields the required equivalence.

6.3. Some Examples.

L. Hörmander introduced a class of operations $L(D)$ with variable coefficients ("of principal type" [33]), having the following property. Let m be the order of $L(D)$, $u \in \mathbb{H}$, and let the equation $\mathbf{L}^{wk} u = f$ hold locally, where $f \in \mathbb{H}$. Then u has (at least locally) all derivatives up to order $m-1$ inclusive.

All this means the following. Let $u \in \mathbb{H}$ have support concentrated in a ball of sufficiently small radius (recall that the support of $u(x)$ is the closure of the set on which $u \neq 0$; for an element $u \in \mathbb{H}$ the definition requires some evident modifications), and for every $v \in C^\infty$ let u satisfy the equation

$$(u, \mathbf{L}^t(D)\, v) = (f, v) \tag{6}$$

for some $f \in \mathbb{H}$. In view of the hypotheses on u, the integration in the definition of the scalar product may be supposed to be extended over the whole space. It is established that then there is a function f^β, belonging to $\mathbb{H}$, such that

$$(u, D^\beta v) = (-1)^{|\beta|} (f^\beta, v)$$

for every $v \in C^\infty$ and every multi-index β for which $|\beta| \le m - 1$.

Using the results of the present subsection, we may immediately conclude that for operations of principal type the following proposition is always valid: if $u \in \mathbb{H}$ and satisfies the hypotheses made above (including equation (6)) then

$$\| \mathbf{L} J_\varepsilon u - f \| \to 0 \quad \text{as } \varepsilon \to 0$$

for every averaging operator J_ε. The integration in the definition of the norm is over the whole space.

Let us give an example to show that the use of equations of the form (3) is not compulsory in proving the equivalence of weak and strong extensions of operations with variable principal part. Here our discussion will have a global character, in distinction from the local point of view used above.

Let V be a unit square in the (x_1, x_2) plane, lying in the first quadrant, and

$$\mathbf{L}(D) = x_2 D_1^2 - D_1 D_2 \, .$$

Let us show that $u \in \mathfrak{D}(\tilde{\mathbf{L}}^{wk})$ implies $u \in \mathfrak{D}(\tilde{\mathbf{L}})$, although the operation does not have principal type in V and in general equations of the form (3) for $u \in \mathfrak{D}(\tilde{\mathbf{L}}^{wk})$ cannot be proved (see [33]).

If $u \in \mathfrak{D}(\tilde{\mathbf{L}}^{wk})$ there is an element $f \in \mathbb{H}$ such that the equation

$$(u, \mathbf{L}^t J\, v) = (f, J\, v)$$

(the parentheses denote the scalar product in $\mathbb{H}(V)$) is satisfied for every element $v \in \mathbb{H}$ if the averaging operator J has the property that it carries v to a function in $C_0^\infty(V)$.

As J we take the operator $\dot{J}_\varepsilon \dot{J}_\eta$, where $\dot{J}_\varepsilon$ is the "contracting" operator discussed in subsection 1, §5, acting on x_1, and $\dot{J}_\eta$ is the same operator (with averaging radius η) acting on x_2. Then

$$(u, \mathbf{L}^t \dot{J}_\varepsilon \dot{J}_\eta v) = (u, [x_2 D_1^2 - D_1 D_2] \dot{J}_\varepsilon \dot{J}_\eta v)$$
$$= (D_1^2 \tilde{J}_\varepsilon \tilde{J}_\eta [x_2 u], v) - (D_1 D_2 \tilde{J}_\varepsilon \tilde{J}_\eta u, v). \tag{7}$$

We take the sequence $\varepsilon_k = 2^{-k}$, $k = 1, 2, \dots$, and write $w_k = D_1^2 \tilde{J}_{\varepsilon_k} u$. Then the first scalar product on the right-hand side of (7) can be written in the form $(\tilde{J}_\eta[x_2 w_k], v)$ (the coefficient x_2 is permutable with differentiation and averaging with respect to x_1). We choose the sequence η_k so that

$$\|\tilde{J}_{\eta_k}(x_2 w_k) - x_2 \tilde{J}_{\eta_k} w_k\| = \|q_k\| \leq 2^{-k},$$

which is possible because of the properties of averaging. Denoting the operator $\tilde{J}_{\varepsilon_k} \tilde{J}_{\eta_k}$ by $\tilde{J}_k$, we have the equations

$$(u, L^t \tilde{J}_k v) = ([x_2 D_1^2 - D_1 D_2] \tilde{J}_k u + q_k, v) = (\tilde{J}_k f, v),$$

valid for every element $v \in \mathbb{H}$. Or,

$$L \tilde{J}_k u = \tilde{J}_k f - q_k, \quad \|q_k\| \to 0 \quad \text{as } k \to \infty,$$

which also proves that u belongs to the domain of $\tilde{L}$.

This example is also interesting because the operation $L(D)$ was suggested by Hörmander as an operation for which global agreement (in a pre-assigned domain V) of weak and strong extensions cannot be established by the use of "standard" averaging operators (see [33]). The method we used of choosing different averaging radii for different variables is often useful (see [22]).

6.4. Equivalence of Weak and Strong Extensions as a Corollary of the Unique Solvability of Problems. Let V be a normal domain, $L(D)$ a differential operation defined in V, and γ a system of boundary conditions for which we can associate the operators L_γ, L_γ^{wk}: $\mathbb{H} \to \mathbb{H}$ with $L(D)$ (see §4). We shall also suppose that we have defined the transposed operation of $L(D)$ and the system of conditions $t\gamma$ adjoint to conditions γ.

The question of the coincidence of L_γ and L_γ^{wk} plays a central role in one of the classical methods of proving that the operator L_γ: $\mathbb{H} \to \mathbb{H}$ is regular or that the equation

$$L_\gamma u = f, \quad f \in \mathbb{H}, \tag{8}$$

has a unique generalized solution for every element $f \in \mathbb{H}$. We present this method.

For an operator L_γ, taken in the strong sense (i.e. as the strong extension of $L(D)$ under conditions γ) let it be possible to establish the "energy inequality"

$$\|u\| \leq c \|L_\gamma u\|. \tag{9}$$

Two propositions follow immediately from (9):
1. *A strong solution of equation* (8) *is unique.*
2. *The range* $\mathfrak{R}(L_\gamma) \subset \mathbb{H}$ *is a closed subspace of* $\mathbb{H}$.

But according to the definition given in §4, the space

$$\mathbb{H} \ominus \Re(\mathbf{L}_\gamma)$$

consists of weak solutions of the equation $\mathbf{L}_{t\gamma}^t v = 0$.

If the strong solutions of the adjoint problem satisfy an inequality analogous to (9):

$$\|v\| \le c \|\mathbf{L}_{t\gamma}^t v\|, \tag{9 t}$$

and the weak solutions are also strong, then it follows immediately from (9) and (9 t), and the preceding discussion, that the operators $\mathbf{L}_\gamma$, $\mathbf{L}_{t\gamma}^t$: $\mathbb{H} \to \mathbb{H}$ are regular.

This method has its widest application in the case when $\mathbf{L}(D)$ is a system of linear differential operations of first order, and $u = (u_1, \ldots, u_N)$. This is connected precisely with the fact that in this situation the use of averaging operators is most convenient (see [29], [34]).

What is now of interest is the possibility of inverting the preceding discussion, in the sense that from (9) and (9 t) and the existence and uniqueness theorem for a strong solution in $\mathbb{H}$ of the corresponding problem with arbitrary right-hand side there automatically follows the coincidence of $\mathbf{L}_\gamma$ and $\mathbf{L}_\gamma^{wk}$. In fact, it follows from the solvability of the equation $\mathbf{L}_{t\gamma}^t v = g$ for every element $g \in \mathbb{H}$ that the equation $\mathbf{L}_\gamma^{wk} w = 0$ has only the zero solution; consequently, the equation $\mathbf{L}_\gamma^{wk} u = f$ is uniquely solvable, i.e. the weak and strong solutions necessarily coincide.

6.5. The Case of an Ordinary Differential Operator. Now let $\mathbf{L}_\gamma$: $\mathbb{H} \to \mathbb{H}$ be the operator generated by an ordinary differential operation of the form (1), §2 $(n = 1)$, and a homogeneous system of boundary conditions γ for which there is defined an adjoint system $t\gamma$. Ordinarily there is no difficulty in showing, for the classical (and consequently for the strong) solutions of the equations

$$\mathbf{L}_\gamma u = f, \qquad \mathbf{L}_{t\gamma}^t v = g,$$

both inequalities (9), (9 t), and theorems on the existence of solutions. We then obtain the following proposition as a corollary of the considerations of the preceding subsection.

Proposition 2. *Under our hypotheses we have the equations*

$$\mathbf{L}_\gamma^{wk} = \mathbf{L}_\gamma, \qquad \mathbf{L}_{t\gamma}^{t\,wk} = \mathbf{L}_{t\gamma}^t. \tag{10}$$

The presence of equations (10) does not in general eliminate the question of an effective construction of a sequence of functions in C^m (satisfying, possibly in the limit, the required boundary conditions) that approximate the weak solution. No regular methods are known for constructing such sequences.

Proposition 2 is not directly applicable to the case when $L_\gamma = \tilde{L}$ (where $\tilde{L}$ is the corresponding maximal operator). However, the use of the special case $n = 1$ allows us to establish the desired equivalence of weak and strong definitions in this case also.

Proposition 3. *If* $L(D)$ *is an ordinary differential operation and the classical theorems on the representation of the general solution of the homogeneous equation* $L(D)u = 0$ *are valid, then* $\tilde{L}^{wk} = \tilde{L}$.

Proof. We observe that since $N(\tilde{L}^{wk})$ (the kernel of $\tilde{L}^{wk}$) can be represented in the form

$$N(\tilde{L}^{wk}) = \mathbb{H} \ominus \mathfrak{R}(L_0^t),$$

where L_0^t is taken in the strong sense (i.e., may be supposed to be taken in the classical sense), then

$$N(\tilde{L}^{wk}) = N(\tilde{L}) = N.$$

If now $u \in \mathfrak{D}(\tilde{L}^{wk})$, $\tilde{L}^{wk} u = f$, and $v \in \mathbb{H}$ is a solution of the equation $\tilde{L}v = f$, then $\tilde{L}^{wk}(v - u) = 0$, i.e. $u = v + \nu$, $\nu \in N$; consequently $u \in \mathfrak{D}(\tilde{L})$. $\square$

Corollary. *Under our hypotheses,* $L_0 = L_0^{wk}$.

In fact, since $\tilde{L}$ and $\tilde{L}^{wk}$ coincide, so do the adjoints of these operators. $\square$

§7. *W* Spaces

7.0. Introductory Remarks. One of the most fundamental questions in the study of the properties of regular operators or of properties of the corresponding generalized solutions of a boundary value problem

$$L(D)u = f, \qquad \gamma u|_S = 0, \tag{1}$$

is the question of under what additional hypotheses on f, $S = \partial V$, γ, and, of course, on the operation $L(D)$ itself, one can assert that a solution of the equation

$$L_\gamma u = f$$

that belongs, a priori, only to $\mathbb{H}(V)$ will be "classical", i.e. will have the derivatives that appear in the equation and in the boundary conditions (and possibly derivatives of higher order), and will satisfy (1) in the ordinary sense.

As we have seen, this question, in spite of its importance, lies outside the basic problems of this book. However, some indications of the methods of investigating it will be useful. A fundamental role in this investigation is played by particular function spaces, the W^k spaces, consisting of functions that possess generalized derivatives up to order k inclusive. This section is devoted to a brief description of these spaces and some of their properties. A basic source of information about these spaces is the monograph [20]. The books [3] and [19] contain general discussions of everything related to this circle of questions.

7.1. Weak and Strong Generalized Derivatives. For the constructions that we need, it is convenient to use the definition given above of various extensions of an arbitrary operation $L(D)$. Suppose that there is given a normal domain $V \subset \mathbb{R}^n$. Let $D^\alpha = D_1^{\alpha_1} \ldots D_n^{\alpha_n}$ be a given monomial. Setting $L(D) = D^\alpha$, we can define the minimal and maximal operators $\tilde{D}^\alpha, D_0^\alpha$.

Definition. An element $u \in \mathbb{H}$ has a weak (or strong) generalized derivative D^α in V if

$$u \in \mathfrak{D}(\tilde{D}_{wk}^\alpha) \quad (\text{or } u \in \mathfrak{D}(\tilde{D}^\alpha)).$$

We denote the corresponding derivatives by $D_{wk}^\alpha u$ and $D^\alpha u$.

To establish the existence of weak derivatives of some element $u \in \mathbb{H}$ is often simpler than to establish the existence of strong derivatives. For example, we have the following test for the existence of a weak derivative.

Proposition 1. *If $u \in \mathbb{H}$ and there is a sequence of smooth functions $u_i \to u$ (convergence in $\mathbb{H}$), such that $\|D^\alpha u_i\| \leq C$, then $D_{wk}^\alpha u$ exists.*

Proof. For every $\varphi \in C_0^\infty(V)$ we have the equation

$$\int\limits_V D^\alpha \varphi \cdot u_i \, dV = (-1)^{|\alpha|} \int\limits_V \varphi D^\alpha u_i \, dV. \tag{2}$$

We obtain what is required by choosing a weakly convergent subsequence $\{D^\alpha u_i'\}$, substituting it into (2), and then taking the limit as $i \to \infty$. □

At the same time, the weak definition of a derivative is not convenient for proving such facts as, for example, the existence of derivatives of lower order when those of higher order exist, the admissibility of multiplication by smooth functions, etc. Hence the following corollary of the results of §6 is very useful.

Proposition 2. *In a normal domain V the weak and strong definitions of the derivative (or of the operators $\tilde{D}_{wk}^\alpha$ and $\tilde{D}^\alpha$) are equivalent.*

7.2. Spaces W^m, $\mathring{W}^m$, and Embedding Theorems. Supposing first that there is given a normal domain $V \subset \mathbb{R}^n$, we define a scalar product for functions $u \in C^m$ by setting

$$\{u, v\}_m = \int\limits_V \Big[\sum_{|\alpha| = m} D^\alpha u \, \bar{D}^\alpha v + u \, \bar{v} \Big] \, dV. \tag{3}$$

The completion of C^m under the norm induced by this scalar product yields a Hilbert space W^m. For the norm in W^m we use the notation $|u, W^m|$:

$$|u, W^m|^2 = \{u, u\}_m.$$

Omission of the term $u\bar{v}$ in (3) would lead us to a seminorm, i.e. to the definition of a space in which the elements that are polynomials of degree less than m have zero norm.

If we take the original linear manifold in which the product (3) is to be introduced to be $C_0^m(V)$, the smooth functions that vanish on the boundary of V, then the term $u\bar{v}$ in (3) may be omitted. The resulting space is usually denoted by $\dot{W}^m$ (or $\mathring{W}^m$). We take the space W^m as an object for further study and present a number of basic facts concerning it. As for proofs, we restrict ourselves to the simplest situations, the case when m and $n \leq 2$, which is enough to make the content of the propositions clear. Adequate expositions of the corresponding constructions are given in almost all modern texts on functional analysis (for example, in [5] and [10]).

Proposition 3. *We have the embedding $\dot{W}^1 \subset \mathbb{H}$, and the elements $u \in \dot{W}^1$ satisfy the inequality*

$$\|u\| \leq c|u, \dot{W}^1|, \tag{4}$$

where the constant C depends only on V. In addition, for every smooth $(n-1)$-dimensional hypersurface $Q \subset V$ the integral $\int_Q u^2 dQ$ is defined, and if Q is the boundary of V then $\int_Q u^2 dQ = 0$.

Proof. Let $V = (0, b_1) \times (0, b_2)$. If $u \in C_0^1$ we may, for example, write the following chain of inequalities:

$$\int_0^{b_1} u^2(x, y)\, dx = \int_0^{b_1} \left(\int_0^y \frac{\partial u}{\partial y'}\, dy' \right)^2 dx \leq \int_0^{b_1} y \int_0^y \left(\frac{\partial u}{\partial y'} \right)^2 dy'' dx \leq y|u, \dot{W}^1|^2. \tag{5}$$

If we integrate the final inequality that follows from (5), with respect to y, between the limits 0 and b_2, we obtain

$$\|u\|^2 \leq \frac{b_2^2}{2}|u, \dot{W}^1|^2,$$

from which (4) follows for elements $u \in C_0^1$. If now $\{u_i\}$ is a sequence of elements of C_0^1, converging to u in the $\dot{W}^1$ norm, it follows from (4) that the sequence also converges in $\mathbb{H}$. Identification of the limit elements yields the required embedding, and inequality (4) remains valid for every element $u \in \dot{W}^1$. Correspondingly, (5) provides the existence of the required integral for $u \in \dot{W}^1$ for every $y \in [0, b_2]$ and the equation $\int_0^{b_1} u^2(x, 0)\, dx = 0$. A generalization of these arguments yields the existence of a suitable integral along any smooth curve in V. □

Proposition 4. *For* $2m>n$ *the elements of* $\dot{W}^m$ *are continuous functions that vanish on the boundary of* V.

Remark. The elements of $\dot{W}^m$, like the elements of $\mathbb{H}$, are equivalence classes. The statement that an element $u \in \dot{W}^m$ is continuous is to be understood in the same sense as the statement that an element of $\mathbb{H}$ is continuous (see § 1).

Proof. Once again, if $u \in C_0^2$ and $V=(0, b_1) \times (0, b_2)$, if we use a chain of inequalities analogous to (5) and the corresponding discussion with approximating sequences, we obtain the existence of the restriction of an arbitrary element $u \in \dot{W}^2$ to an arbitrary curve $l \in V$ and the fact that this element belongs to the space $\dot{W}^1(l)$ (the ends of the curve lie on $\partial V'$). If, however, $u \in \dot{W}^1$, for example on $(0, b)$, we may use the representation

$$u_i(x) = \int_0^x \frac{\partial u_i}{\partial \xi} \, d\xi,$$

for the approximating sequence, and an obvious argument (for example, estimating the difference $|u(x_1)-u(x_2)|$) lets us establish the continuity of the corresponding element $u \in \dot{W}^1$. It is clear that for a continuous function the condition of vanishing on the boundary of V is satisfied in the ordinary sense.

Remark. It is easy to see, using only simple arguments, that when $n=3$, in order to establish the continuity of $u \in \dot{W}^m$ we would succeed with $m=3$, whereas in fact $m=2$ suffices. To obtain precise results we would need more complicated machinery.

Proposition 4 (and its variants) is the fundamental tool for establishing the smoothness of generalized solutions, as was mentioned in the introduction. Having shown by some auxiliary construction that a generalized solution belongs to some space of the type of W^m for sufficiently large m, we can draw conclusions about its smoothness in the classical sense.

Proposition 5. *The embedding* $\dot{W}^{k+1} \subset \dot{W}^k$ *is completely continuous.*

The conclusion means that every set that is bounded in $\dot{W}^{k+1}$ is compact in $\dot{W}^k$. For the proof it is enough to consider the embedding of $\dot{W}^1$ in $\mathbb{H}$. We present the standard method. The boundedness of a set $\mathcal{M} \subset \dot{W}^1$ implies its boundedness in $\mathbb{H}$ and equicontinuity in the mean (cf. § 1). Hence it follows that for every given $\varepsilon > 0$ the set $\{J_\varepsilon f\}$, $f \in \mathcal{M}$ (J_ε is an averaging operator) is equicontinuous in the classical sense in $C(V)$. It is consequently compact (Arzelà's theorem), and for every $\varepsilon_1 > 0$ there is a finite ε_1-net in C, which implies the existence of a corresponding net in $\mathbb{H}$. □

We content ourselves with this brief exposition of the simplest facts concerning the theory of function spaces whose elements have generalized derivatives.

Chapter III
Ordinary Differential Operators

§0. Introductory Remarks

It is natural to begin the study of those aspects of the theory of boundary value problems that are of interest here with a rather detailed discussion of a number of properties of ordinary differential equations. In the first place, ordinary differential operations are the simplest entities in the theory in which we are interested that can, in many respects, be studied exhaustively. In the second place, the fundamental kind of operators that are of interest in what follows will be operators that act on tensor products $\mathbb{H}_0 \otimes \mathbb{H}_1 \otimes \ldots \otimes \mathbb{H}_n$ (Chapter IV) and are constructed from mutually commuting operators $\mathbf{L}_k : \mathbb{H}_k \to \mathbb{H}_k$. Here the material for the construction of $\mathbf{L}_k$ will be ordinary differential operators that have properties depending on the boundary conditions that enter into their definitions.

We should observe that from the point of view of the theory of boundary value problems for partial differential equations, the entities described above are models, i.e. very much simplified in comparison with the basic general entities, namely arbitrary linear differential operations considered in a bounded domain in $\mathbb{R}^n$. However, we hope to show that these models are worth studying and let us analyze many phenomena that arise in the theory of boundary value problems.

A third advantage of making a detailed study of ordinary differential operations (one that is closely connected with the second) is that many particular phenomena that arise in the study of partial differential operations can be conveniently investigated in terms of transfer from an ordinary differential operation (for example, from the operation $\mathbf{L}(\mathbf{A}, D) = \mathbf{A}_2 D^2 + \mathbf{A}_1 D + \mathbf{A}_0$, where the $\mathbf{A}_k$ are numbers) to a corresponding "equation with operator coefficients" or an "equation in a Banach space" (i.e., to an operation $\mathbf{L}(\mathbf{A}, D)$ where $\mathbf{A}_k$ are operators). In order to effect such a transfer it is indispensable to consider attentively a number of properties of ordinary differential operations.

§1. Description of Proper Operators for $n = 1$

As an example of an interesting problem which can be completely solved in the case of an ordinary differential operation, we present a full description of the proper operators when V is a finite interval on the real line.

1.1. Operators Generated by Cauchy Conditions. Let

$$x \in V = (0, b), \quad \mathbb{H} = \mathbb{H}(V), \quad D \equiv \frac{d}{dx}, \quad L(D) = \sum_{0}^{m} a_k(x) D^k \tag{1}$$

and let L_c: $\mathbb{H} \to \mathbb{H}$ be the operator defined as the closure in $\mathbb{H}$ of the operation (1), considered initially on smooth functions that satisfy the Cauchy conditions

$$u|_0 = u'|_0 = \ldots = u^{(m-1)}|_0 = 0.$$

We shall assume from now on that $a_k(x)$ is continuous in $\bar{V}$ and $a_m \equiv 1$.

Theorem 1. *The operator* L_c^{-1}: $\mathbb{H} \to \mathbb{H}$ *exists, and is a Volterra operator.*

Remark. The information about the solution of the equation $L_c u = f$ that we obtain in the course of the proof is far from being exhausted by the conclusion of the theorem. We carry out the proof by a method that can be generalized to hyperbolic partial differential operators (see [7]).

Lemma 1. *Let the nonnegative integrable function* $\rho(x)$, $x \in (0, b)$, *satisfy the inequality*

$$\rho(x) \leq c \int_0^x \rho(\xi) \, d\xi + \sigma(x), \quad c = \text{const} > 0, \tag{2}$$

where $\sigma(x) \geq 0$ *is a bounded nondecreasing function of x. Then*

$$\rho(x) \leq e^{cx} \sigma(x). \tag{3}$$

Proof. We introduce the notation $J\varphi = \int_0^x \varphi(\xi) \, d\xi$. Then, writing (2) in the form $\rho \leq cI\rho + \sigma$ and applying the operation cI to this inequality, we obtain

$$cI\rho \leq c^2 I^2 \rho + cI\sigma.$$

Inserting this inequality for $cI\rho$ into (2), we have

$$\rho \leq c^2 I^2 \rho + cI\sigma + \sigma. \tag{4}$$

We may again apply the operation cI to (4) and again insert the result into (2). Repeating this process n times, we obtain

$$\rho \leq c^{n+1} J^{n+1} \rho + c^n J^n \sigma + c^{n-1} J^{n-1} \sigma + \ldots + \sigma. \tag{5}$$

Now observe that

$$J^n \sigma(t) = \int_0^t \frac{(t-\xi)^{n-1}}{(n-1)!} \sigma(\xi) \, d\xi \le \frac{t^n}{n!} \sigma(t) \tag{6}$$

($\sigma(t)$ is nondecreasing) and that

$$\varepsilon_n = c^{n+1} J^{n+1} \rho(t) \le c^{n+1} \frac{b^n}{n!} \int_0^b \rho(t) \, dt \tag{7}$$

tends to zero as $n \to \infty$. Substituting (6) and (7) into (5) and taking the limit as $n \to \infty$, we obtain (3). $\quad\square$

Lemma 2. *If* $u \in \mathfrak{D}(L_c)$, $m \ge 1$, $L_c u = f \in \mathbb{H}$, *then*

$$|D^{m-1} u(x)|^2 \le e^{cx} \int_0^x |f|^2 \, d\xi, \tag{8}$$

where the constant c *depends only on* m *and the coefficients* a_k *of* (1).

Proof. It is evidently enough to carry out the proof of (8) for smooth functions $u(x)$ that satisfy the Cauchy conditions. Multiply the equation

$$L(D) u(x) = f(x)$$

by $D^{m-1} u$ and integrate between the limits 0 and x. We obtain

$$\int_0^x D^m u \, D^{m-1} u \, d\xi + \sum_{0}^{m-1} \int_0^x a_k(\xi) D^k u \, D^{m-1} u \, d\xi = \int_0^x f(\xi) D^{m-1} u \, d\xi.$$

If we now use the equation

$$\left| \int_0^x D^m u \, D^{m-1} u \, d\xi \right| = \frac{1}{2} \left| \int_0^x D(D^{m-1} u)^2 \, d\xi \right| = \frac{1}{2} |D^{m-1} u(x)|^2$$

and the inequalities

$$\max_{k,x} |a_k(x)| \le M, \quad |D^k u \, D^{m-1} u| \le \tfrac{1}{2}(|D^k u|^2 + |D^{m-1} u|^2),$$

$$|f(\xi) D^{m-1} u| \le \tfrac{1}{2}(|f|^2 + |D^{m-1} u|^2),$$

as well as the fact that for every $k \ge 0$

$$\int_0^x |D^k u|^2 \, d\xi = \int_0^x \left| \int_0^\xi D^{k+1} u \, d\eta \right|^2 d\xi \le \int_0^x \xi \int_0^\xi |D^{k+1} u|^2 \, d\eta \, d\xi \le \frac{x^2}{2} \int_0^x |D^{k+1} u|^2 \, d\xi,$$

we obtain

$$|D^{m-1}u(x)|^2 \leq C\int_0^x |D^{m-1}u|^2 \, d\xi + \int_0^x |f|^2 \, d\xi.$$

From this we obtain the conclusion of Lemma 2 by applying Lemma 1. $\square$

Corollary. *Under the hypotheses of Lemma 2 we have the inequality*

$$\|u\| \leq c \|\mathbf{L}_c u\|. \tag{Φ}$$

Inequality (Φ) is evidently a crude version of (8).

Lemma 3. *All points* $\lambda \in \mathbb{C}$ $(|\lambda| < \infty)$ *belong to the resolvent set of* $\mathbf{L}_c$.

First proof. It follows immediately from (Φ) that for every λ (which may be included in a_0) there exists a bounded operator $(\mathbf{L}_c - \lambda)^{-1}$. It is only necessary to verify that the domain of this operator is the whole space $\mathbb{H}$, i.e. that the equation $\mathbf{L}_{c,\lambda}u = f$ (or simply $\mathbf{L}_c u = f$) is solvable for every $f \in \mathbb{H}$. This follows at once from the classical theory of ordinary differential equations (for example, for $f \in C$, which is enough because of (Φ)). $\square$

Second proof. If we wish to obtain a proof independently of classical arguments, as is necessary for partial differential operators, we can use the method explained in 6.4, Chapter II.

The orthogonal complement of $\mathfrak{R}(\mathbf{L}_c)$, the range of $\mathbf{L}_c$, is, by (Φ), the closed subspace of $\mathbb{H}$ consisting of the weak solutions of the equation

$$\mathbf{L}_{tc}^t v = 0, \tag{9}$$

where $t c$ are the Cauchy conditions for $x = b$. The proof will be complete if we show that (9) implies $v = 0$. A strong solution of (9) is necessarily identically zero (because we have the corresponding inequality (Φ) for $\mathbf{L}^t$ and conditions $t c$, as is evident). However, to assert the identity of the weak and strong solutions (§6 of Chapter II), we need to require that $a_k(x) \in C^k$. $\square$

We may also observe that in proving the equivalence of the weak and strong definitions of the operator $\tilde{\mathbf{L}}$ for ordinary differential operations we used the finite-dimensionality of the kernel $\tilde{\mathbf{L}}$, which is implicit in the corresponding classical result. But this result follows immediately from the uniqueness of the solution of the Cauchy problem, which was established above (independently).

The Volterra character of $\mathbf{L}_c^{-1}$, i.e. the conclusion of Theorem 1, follows immediately from Lemma 3 and the lemma on the connection between the spectra of $\mathbf{T}$ and $\mathbf{T}^{-1}$ (2.3, Chapter I).

We also notice that the constant c in (Φ) that appears in the operator $\mathbf{L}_c - \lambda$ also evidently depends on λ. The question of the nature of this dependence is a question about the behavior of the resolvent of $\mathbf{L}_c$. In what follows we shall frequently need to carry out a corresponding investigation of the resolvent, usually also in connection with the investigation of the spectral properties of the operator (understood in the extended sense).

For an ordinary differential operator, inequality (8) and its corollary

$$\|D^{m-1}u\| \le c\|f\|$$

immediately imply the corresponding inequality for arbitrary order m.

Lemma 4. *If* $u \in \mathfrak{D}(L_c)$, $m \ge 1$, $L_c u = f \in \mathbb{H}$, *then*

$$\|D^m u\| \le c\|f\|,$$

where c is a constant independent of f.

Proof. Multiply the equation $L(D)u = f$ by $D^m u$ and integrate from 0 to b; using the boundedness of $a_k(x)$ and the Cauchy-Bunyakovsky inequality, we immediately obtain

$$\|D^m u\|^2 - c\|D^m u\| \sum_0^{m-1} \|D^k u\| \le \|f\| \, \|D^m u\|.$$

Dividing by $\|D^m u\|$ and using the fact that the norms of the derivatives $D^k u$, $k \le m-1$, are bounded by the norm of f (cf. the proof of Lemma 2), we obtain the conclusion. $\square$

1.2. Description of Proper Operators. First let $V = (0, b)$, $\mathbb{H} = \mathbb{H}(V)$, and let $L(D)$ let the ordinary differential operation (1). We use the terminology of Chapter II.

Lemma 5. *If* **L** *is a proper restriction of the maximal operator* $\tilde{\mathbf{L}}$, *then* $\mathbf{L}^{-1}$ *admits the representation*

$$\mathbf{L}^{-1}f = \mathbf{L}_c^{-1}f + \sum_0^{m-1} l_k(f)\,\omega_k, \tag{10}$$

where f is any element of $\mathbb{H}$, $\mathbf{L}_c$ is the operator generated by the Cauchy conditions, l_k is a bounded linear functional on $\mathbb{H}$, and $\{\omega_k\}_0^{m-1}$ is a given fundamental system of solutions of the equation $L(D)w = 0$.

Proof. Since the kernel of $\tilde{\mathbf{L}}$ is finite-dimensional, it follows at once that for every element $u \in \mathfrak{D}(\tilde{\mathbf{L}})$ the strong derivatives $u^{(k)}$, $k \le m$, exist, and consequently every solution of the homogeneous equation $\mathbf{L}u = f$ is representable, as in the classical case, as a sum of a particular solution and the general solution of the homogeneous equation, i.e. in the form of the right-hand side of (10) with some constants l_k. The choice of the constants will depend, generally speaking, both on f and on the chosen restriction of $\tilde{\mathbf{L}}$ (in the classical case, on the corresponding boundary conditions). The dependence on f is linear because $\mathbf{L}$ is linear; the corresponding functionals must be bounded since $\mathbf{L}^{-1}$ is bounded and the system of solutions $\{\omega_k\}$ are independent. $\square$

Lemma 6. *For a given proper restriction $\mathbf{L}$, the domain $D(\mathbf{L})$ consists of the elements $u \in \mathfrak{D}(\tilde{\mathbf{L}})$ that satisfy the supplementary system of conditions*

$$u^{(j)}|_0 = \sum_0^{m-1} (q_k, \tilde{\mathbf{L}}u) \, \omega_k^{(j)}(0), \quad j = 0, 1, \ldots, m-1. \tag{11}$$

Here the parentheses denote the scalar product in $\mathbb{H}$, and $\{q_k\}_0^{m-1}$ is the set of elements of $\mathbb{H}$ that determine the proper restriction.

Proof. Every element $u \in \mathfrak{D}(\tilde{\mathbf{L}})$ has, as we noted, m generalized derivatives, and consequently both the boundary values of u and of their derivatives that appear in (11) are determined (cf. §7 of Chapter II). If we replace $\mathbf{L}^{-1}f$ by u on the left-hand side of (10), we obtain the functionals $l_k(f)$ as scalar products (q_k, f) (Riesz's lemma), if we observe that the values of $\mathbf{L}_c^{-1}f$, and of the derivatives up to order $m-1$, are zero at zero; replacing f by $\mathbf{L}^{-1}f$ we return to the system of conditions (11). $\square$

Lemma 7. *A proper restriction $\mathbf{L} \subset \tilde{\mathbf{L}}$ is an extension of the corresponding minimal operator $\mathbf{L}_0$ if and only if the elements $\{q_k\}$ in (11) belong to the kernel of $\tilde{\mathbf{L}}^!$.*

Proof. Necessity. Let $\mathbf{L}_0 \subset \mathbf{L}$. Then formula (11) holds for every $v \in \mathfrak{D}(\mathbf{L}_0)$, and $v^{(j)}|_0 = 0$ for all values of j that appear in (11). But then, since the $\omega_k^{(j)}$ are independent, it follows that $(q_k, \mathbf{L}_0 v) = 0$ for all k and every $v \in \mathfrak{D}(\mathbf{L}_0)$. But this means that q_k belongs to the kernel of $\tilde{\mathbf{L}}^!$ because of the equivalence of weak and strong extensions for ordinary differential operators.

Sufficiency. If the elements $\{q_k\}$ in (11) belong to the kernel of $\tilde{\mathbf{L}}^!$ then they have, as we noticed, m strong derivatives, and we may interchange $\tilde{\mathbf{L}}$ and q_k in the scalar products $(q_k, \tilde{\mathbf{L}}u)$ by means of integration by parts. Since $\tilde{\mathbf{L}}^!q_k = 0$, conditions (11) take the form of homogeneous boundary conditions

$$u^{(j)}|_0 - \sum_0^{m-1} \mathscr{L}_k^j(u)|_0^b = 0, \quad j = 0, 1, \ldots, m-1. \tag{12}$$

Each sum in (12) is a linear combination of the values of u and its derivatives up to order $m-1$, which reduce to zero at b. The coefficients of these linear combinations depend on $\{q_k\}$, $\{\omega_k\}$, and the coefficients of $\mathbf{L}(D)$. With (11) presented in the form (12), the embedding $\mathbf{L}_0 \subset \mathbf{L}$ follows. $\square$

The preceding discussion allows us to state, in particular, the following theorem.

Theorem 2. *Under our hypotheses every proper operator $\mathbf{L}$ that is generated by an ordinary differential operation $\mathbf{L}(D)$ is specified by giving the domain*

$\mathfrak{D}(L)$ *together with the boundary conditions* (12) *which an element of* $\mathfrak{D}(\tilde{L})$ *is to satisfy.*

It follows from the discussion in subsection 1 that when $n=1$ the linear manifold $\mathfrak{D}(\tilde{L})$ coincides with the space $W^m(V)$.

§2. The Ordinary Differential Operator of the First Order

An ordinary operation of the first order is ordinary *par excellence*. We shall discuss it in great detail, having in mind, on the one hand, to illustrate the results of §1 by the simplest examples, and on the other, to prepare for the study of the corresponding operator equations.

As in §1, we suppose that $x \in V = (0, b)$, $\mathbb{H} = \mathbb{H}(V)$, and that $L(D)$ has the form

$$L(D) \equiv D - a(x), \qquad D \equiv \frac{d}{dx}, \tag{1}$$

where $a(x)$ is a continuous function (supposed, for simplicity, to be real). Formula (10) of §1 then assumes the form

$$L^{-1} f = \int_0^x \exp \left\{ \int_\xi^x a(\eta)\, d\eta \right\} f(\xi)\, d\xi + l(f) \exp \left\{ \int_0^x a(\xi)\, d\xi \right\}, \tag{2}$$

where $l(f)$ is a bounded functional on $\mathbb{H}$. If $l(f) = (q, f)$, $q \in \mathbb{H}$, then (11) of §1 can be written in the form

$$u|_0 - (q, \tilde{L} u) = 0. \tag{3}$$

Conditions (3), which describe the domain of the proper restriction L of the operation $L(D)$ will be boundary conditions if $q(x)$ satisfies the equation $(D + a(x))\, q = 0$, i.e.

$$q = p \exp \left(-\int_0^x a(\xi)\, d\xi \right),$$

where p is a constant. Then condition (12) of §1 can be written as the equation

$$(1 + p)\, u|_0 - p \exp \left\{ -\int_0^b a\, d\xi \right\} u|_b = 0. \tag{4}$$

For an arbitrary complex parameter p the boundary conditions (4) yield a description of all the proper operators generated by $L(D)$. From a somewhat different point of view, formula (4) yields all boundary conditions for which zero is not a point of the spectrum for the operator $L_{(p)} \colon \mathbb{H} \to \mathbb{H}$ generated by the operation $L(D)$.

Remark 1. The preceding property is reflected in the fact that in distinction from the situation that arises when the boundary conditions are given by the equation

$$\mu u|_0 - \nu u|_b = 0 \tag{5}$$

(where μ and ν are complex constants), under conditions (4) the formula (2) which gives a solution of the problem (in which we still have to calculate the value of the constant $l(f)$) does not contain a denominator which can become zero for certain values of p (formula (8), given below, which yields the value of $l(f)$ under conditions (5), does contain such a denominator).

The boundary conditions (4) show that all proper operators generated by the operation (1) are necessarily described by conditions of the form (5) with arbitrary μ and ν. Or, in order to have a single parameter, by the conditions

$$\mu u|_0 - u|_b = 0 \tag{6}$$

with

$$u|_0 = 0, \tag{7}$$

corresponding to the value $\mu = \infty$ in (6). In the simplest case that we have been considering it turns out to be possible to give a complete description of the structure of all proper operators generated by the operation (1).

Theorem. *All proper operators generated by the operation (1) are described by conditions (6) and (7). The operators generated by (6) with $\mu = 0$ and by (7) are V-operators; those generated by (6) with $\mu \neq 0$ are M-operators (cf. 3.5, Chapter I).*

Proof. We have already verified the sufficiency of (6) for describing all proper operators. The part of the conclusion that refers to the Cauchy conditions at the ends of the interval V follows from the results of §1. It remains only to elucidate the spectral properties of operators generated by (6) with $\mu \neq 0$.

If we solve the equation $(\mathbf{L}(D) - \lambda) u = f$ under conditions (6), we obtain the following formula for the constant $c = l(f)$ in (2) (with $a(\xi)$ replaced by $a(\xi) + \lambda$):

$$c = \frac{If}{\mu - \exp\left[\int\limits_0^b (a(\xi) + \lambda)\, d\xi\right]}, \tag{8}$$

where I is an integral operator, defined over $\mathbb{H}$, whose form is not of interest for the time being. It follows from (8) that the values of λ for which the denominator is not zero belong to the resolvent set $\rho \mathbf{L}$. The values of λ for which the denominator is zero are

$$\lambda_k = b^{-1}(\ln|\tilde{\mu}| + i \arg \tilde{\mu} + 2k\pi i), \quad k = 0, \pm 1, \pm 2, \ldots, \tag{9}$$

where $\tilde{\mu} = \mu \exp \int_a^b a(\xi) d\xi$. For $\lambda = \lambda_k$ the corresponding homogeneous equation has the nontrivial solution

$$u_k(x) = \exp \left(\int_0^x a(\xi) d\xi + \lambda_k x \right). \tag{10}$$

This is the eigenfunction corresponding to the eigenvalue λ_k, i.e. $\lambda_k \in P\sigma L$. The set of eigenfunctions (10) is a Riesz basis in $\mathbb{H}$. In fact, this set differs from the complete orthonormal system $\varphi_k = \exp \dfrac{2k\pi i}{b} x$ only by an inessential factor (which does not take either of the values 0, ∞). $\square$

Remark 2. It is classical that the system $\{\varphi_k\}$ introduced above is complete in $\mathbb{H}$. However, this fact can be obtained without difficulty from the results of spectral theory that we have already presented. Indeed, let $L(D) = iD_x - a$ and let the corresponding operator $L: \mathbb{H} \to \mathbb{H}$ be defined by (6) with $\mu = 1$ (periodicity). Then when $a \notin P\sigma L$ the operator L^{-1} (a real) is a self-adjoint CC operator, as follows immediately from, for example, the explicit formula that yields the solution of the equation $Lu = f$ (one could also use the results of §7, Chapter II). Here every smooth periodic function belongs to the range of L^{-1} (the equation $L^{-1}f$ is trivially solvable if $u \in \mathfrak{D}(L)$), and consequently can be expanded (3.3 of Chapter I) in terms of the eigenfunctions $\{\varphi_k\}$ of L^{-1} (of L; see 2.3 of Chapter I). Since the set of smooth periodic functions is dense in $\mathbb{H}$, the system $\{\varphi_k\}$ is complete in $\mathbb{H}$.

Remark 3. An unexpectedly complex question (even for the simplest operation (1)) is the question of the spectrum of a proper restriction of $\tilde{L}$ that is not an extension of the minimal restriction, i.e. of the spectrum of the operator generated by (1) and (3) when $q \notin N(\tilde{L}')$. For a smooth function q we may write (3) in the form

$$(1+q) u|_0 - q u|_b + \int_0^b u \tilde{L} q \, dx = 0;$$

it would be interesting to clarify the nature of the spectra of the corresponding operators.

Formulas (9) and (10) show that the presence in $L(D)$ of the "minor" term $a(x)$ induces only a shift of the spectrum and a change in the structure of the eigenfunctions; this will be unimportant in the majority of the cases that will be of interest in later constructions, and we may restrict ourselves to the case $a(x) = 0$. In this case the explicit representation of the operator L_λ^{-1} generated by (6) will be

$$L_\lambda^{-1} f = \frac{\mu \int_0^x e^{(x-\xi)\lambda} f(\xi) d\xi + e^{b\lambda} \int_x^b e^{(x-\xi)\lambda} f(\xi) d\xi}{\mu - e^{b\lambda}}. \tag{11}$$

§3. Birkhoff Theory

Unfortunately, when $m \geq 2$ the general operation of the form (1), §1, does not admit as exhaustive a study as was carried out in §2 for the case $m = 1$. This occurs even for operations with constant coefficients, when it is possible to give an explicit solution of the equation

$$\mathbf{L}(D) u \equiv \sum_{0}^{m} a_k D^k u = f. \tag{1}$$

In our investigation we intend, first of all, to elucidate the dependence of the spectrum (properties of the resolvent operator) on the choice of the boundary conditions. For an operator of the first order essentially all boundary conditions fill out a family that depends on the single parameter μ, and the values of this parameter determine the position of a vertical line in the complex plane that contains all the points of the spectrum. On the other hand, a satisfactory answer to the question: "How are we to describe the spectra of the operators $\mathbf{L}$ generated by the operation (1) under all possible choices of the boundary conditions?" has not received an answer, as we shall see below, even for $m = 2$.

The nature of the difficulties that arise is made clear by the presence, in the formula for the solution of equation (1), of a denominator in the form of a determinant Δ (playing the role of the function $\mu - \exp b \lambda$ in formula (11) of §2), for which we have to investigate the dependence of its properties (in particular, the distribution of the zeros) on a large number of parameters that enter into boundary conditions of a general form.

The methods for overcoming these difficulties include a number of classical results that basically go back to Birkhoff. The fundamental idea consists of a description of classes of boundary conditions that allows one to separate off a "principal part" of the determinant mentioned above, and to use the asymptotic behavior of its zeros that determine the point spectrum of the corresponding operator. We shall illustrate Birkhoff's construction for the simplest example of a second-order operation. A detailed exposition of the corresponding general procedures is given in [18].

When $m = 2$ the change of variable $v = u \exp \left(-\tfrac{1}{2} \int_0^x a_1(\xi) d\xi \right)$ transforms the operation (1) into the form

$$D^2 u + a(x) u \tag{2}$$

(for a general operation (1) a corresponding transformation will remove the term $a_{m-1} D^{m-1}$).

Proposition 1. *When $|\rho| \to \infty$ the linearly independent solutions of the equation*

$$D^2 u + a(x) u + \rho^2 u = 0 \tag{3}$$

can be represented in the form

$$u_{1,2} = e^{\pm i \rho x}[1 + O(1/\rho)]. \quad \square$$

Remark. For a general operation (1) there is a similar proposition: the solutions of the equation $(L(D) + \rho^m) u = 0$ tend, as $|\rho| \to \infty$, to the functions $e^{\rho \omega_k x}$, $k = 1, \ldots, m$, where ω_k are a family of roots of unity; i.e., they tend to a fundamental system of solutions of the equation $(D^m + \rho^m) u = 0$.

For $x \in V = (0, b)$, we present the general system of boundary conditions for the operation (2):

$$
\begin{aligned}
\Gamma_1(u) &= \alpha_1 u'|_0 + \alpha_{10} u|_0 + \beta_1 u'|_b + \beta_{10} u|_b = 0, \\
\Gamma_2(u) &= \alpha_2 u'|_0 + \alpha_{20} u|_0 + \beta_2 u'|_b + \beta_{20} u|_b = 0.
\end{aligned} \tag{Γ}
$$

If now $c_1 u_{1,\rho} + c_2 u_{2,\rho}$ is the general solution of the homogeneous equation (3), and

$$c_1 u_{1,\rho} + c_2 u_{2,\rho} + u_\rho(f)$$

is the general solution of the nonhomogeneous equation, then in order to obtain a solution satisfying conditions (Γ) we need to solve the system of equations that determine the values of c_1 and c_2. The determinant of this system has the form

$$\Delta(\rho, \Gamma) = \begin{vmatrix} \Gamma_1(u_{1,\rho}) & \Gamma_1(u_{2,\rho}) \\ \Gamma_2(u_{1,\rho}) & \Gamma_2(u_{2,\rho}) \end{vmatrix}.$$

Proposition 2. *The values of ρ for which $\Delta(\rho, \Gamma) = 0$ are the eigenvalues of the operation (2) under conditions (Γ); i.e., for these values of ρ equation (3) has a nontrivial solution that satisfies conditions (Γ).* $\square$

We now set $u_{\sigma,\rho} = e^{\rho \sigma x}$, $\sigma = 1, 2$, and discuss the structure of the determinants Δ. We have

$$
\begin{aligned}
\Gamma_1(u_{1,\rho}) &= \alpha_1 \rho_1 + \alpha_{10} + \beta_1 \rho_1 e^{b \rho_1} + \beta_{10} e^{b \rho_1} \\
&= \rho_1 \{\alpha_1 + O(1/\rho_1) + e^{b \rho_1}[\beta_1 + O(1/\rho_1)]\}.
\end{aligned}
$$

Similarly

$$\Gamma_2(u_{1,\rho}) = \rho_1 \{\alpha_2 + O(1/\rho_1) + e^{b \rho_1}[\beta_2 + O(1/\rho_1)]\},$$

and the corresponding representations for $\Gamma_\sigma(u_{2,\rho})$, $\sigma = 1, 2$, are obtained by replacing ρ_1 by ρ_2.

Proposition 3. *If the condition*

$$\alpha_1 \beta_2 - \alpha_2 \beta_1 \neq 0 \tag{4}$$

is satisfied, *the distribution of the zeros of* $\Delta(\rho, \Gamma)$, *considered as a function of* ρ, *is approximately described by the equation*

$$e^{b(\rho_1 - \rho_2)} = 1. \tag{5}$$

Equation (5) was evidently obtained simply by setting the principal part of Δ equal to zero. A rigorous proof of the location of the corresponding zeros of Δ depends on Routh's theorem (see [18]). $\square$

If we notice that $\rho_2 = -\rho_1 = \rho$, we obtain $\rho = b^{-1} k \pi i$, $k = 0, \pm 1, \pm 2, \ldots$, i.e. the set of eigenvalues is asymptotically the same as for the corresponding problem

$$u|_0 = u|_b = 0$$

for the simplest operation D^2.

Condition (4) is not a necessary condition for being able to isolate the principal part of Δ. Birkhoff's theory introduces the concept of *normalized* boundary conditions (when $\alpha_1 \beta_2 - \alpha_2 \beta_1 = 0$ the boundary conditions (Γ) become normalized after excluding the derivatives u'_0 and u'_b from one of the rows) and defines the class of *regular* conditions that allow a description of the asymptotics of the eigenvalues and eigenfunctions. The corresponding construction serves to select the dominant exponential functions from the set $e^{\rho \omega_k x}$, $k = 1, \ldots, m$, that approximate the fundamental system of solutions of the equation $(\mathbf{L}(D) + \rho^m) u = 0$ (see the remark on Proposition 1).

Afterwards one can introduce the concept of *strongly regular* boundary conditions, under which one can show that the eigenfunctions of the corresponding problem form a Riesz basis in $\mathbb{H}(V)$ (see [37]).

§4. Supplementary Remarks

4.1. General Remarks. The content of these general remarks is really more closely related to the general considerations in Chapter II. However, it is more convenient (although less logical) to locate them after the analysis of the one-dimensional case.

Let $V \subset \mathbb{R}^n$ be a normal domain and let $\mathbf{L}(D)$ be a general differential operation of the form (1), §2, Chapter II.

Definition. An operator $\mathbf{L}_\gamma: \mathbb{H}(V) \to \mathbb{H}(V)$, $\mathbf{L}_0 \subset \mathbf{L}_\gamma \subset \tilde{\mathbf{L}}$, generated by the operation $\mathbf{L}(D)$ and a certain system of boundary conditions γ, is said to be *proper in the wide sense* if its resolvent set $\rho \mathbf{L}_\gamma \subset C$ is non-empty.

An operator that is proper in the sense defined in 3.3, Chapter II, will sometimes be called *proper in the strict sense*.

Boundary conditions that determine an operator $\mathbf{L}_\gamma$ that is proper in the wide sense are "reasonable" in the sense that they are neither overdetermined nor incomplete, and the lack of being proper (unique solvability for an arbitrary right-hand side) of the corresponding equation is connected, as a rule, with the presence of a point spectrum and a "singularity" in the defining parameters of the problem. A large number of interesting theorems that describe properties of one or another specific operator generated by boundary value problems for differential operations include statements that they are proper in the wide sense and propositions on the sets of singular values of the parameters.

In connection with our discussion it is natural to introduce the terminology that is often used in studying the solvability of operator equations both in Hilbert and in Banach spaces (see [13]). An equation

$$\mathbf{T}u = f, \quad \mathbf{T}: \mathbb{H} \to \mathbb{H}, \tag{1}$$

containing a closed operator $\mathbf{T}: \mathbb{H} \to \mathbb{H}$ is said to be *normally-solvable* if $\mathfrak{R}(\mathbf{T})$ is a closed subspace of $\mathbb{H}$; an operator $\mathbf{T}$ is *Noetherian* if the linear manifold $N(\mathbf{T})$ and the subspace $\mathcal{Q} = \mathbb{H} \ominus \mathfrak{R}(\mathbf{T})$ are both finite-dimensional, and is a *Fredholm operator* if the dimensions are equal (in the Western literature Noetherian operators are often called Fredholm operators). The dimension of $\mathcal{Q}$ is the *deficiency* of $\mathbf{T}$; the difference between the dimensions of N and $\mathcal{Q}$ is the *index*.

Consequently an equation (1) in which $\mathbf{T}$ is a minimal operator is, by the propositions of §3 of Chapter II, normally-solvable, but in this case the deficiency of $\mathbf{T}$ is infinite if $n > 1$. If $\mathbf{T}$ is a CC operator then $\mathbf{T}_\lambda$ is a Fredholm operator for every $\lambda \neq 0$; we shall not encounter operators with nonzero index, whose investigation is an important part of the theory of boundary value problems for elliptic differential operations. If λ is a point of the continuous spectrum of $\mathbf{T}$, the equation $\mathbf{T}_\lambda u = f$ is not normally-solvable.

The preceding terminology is used most often when the operator $\mathbf{T}$ in (1) acts on a B-space, i.e. $\mathbf{T}: \mathscr{B}_1 \to \mathscr{B}_2$. In our approach, which is restricted to the case $\mathbf{T}: \mathbb{H} \to \mathbb{H}$, properties based on the definitions of spectral theory provide more complete information.

4.2. Supplementary Remarks Concerning Ordinary Differential Operations. We return to our model example $L(D) = D_x$, $x \in (0, b)$. We remarked in §2 that in this case all the proper operators generated by $L(D)$ are included among the operators given by (6) and (7) of §2. Here the operator is proper if $\mu \neq 1$ in (6), and proper in the wide sense if $\mu = 1$. It is not without interest to arrive at the same result by more complicated considerations.

Let the operator $\mathbf{L}$ (generated by D_x) be proper in the wide sense and let $\lambda \in \rho \mathbf{L}$. On the basis of (4), §2, we can say that the boundary conditions that generate it are among the conditions that we can write in the form

$$(1 + p)u|_0 - p e^{-b\lambda} u|_b = 0. \tag{2}$$

But are not these conditions already contained among the conditions for λ $=0$? If "yes", then

$$1+q=k(1+p), \quad q=kpe^{-b\lambda}. \tag{3}$$

The only cases when equations (3) do not make it possible to find the necessary values of k and q are connected with the possibility of having the equation

$$1+p=pe^{-b\lambda};$$

when this is satisfied, (2) becomes

$$u|_0-u|_b=0.$$

As a result we obtain a somewhat different version of the theorem in §2.

Theorem. *All proper operators generated by the operation D_x on $(0,b)$ are described by the conditions*

$$(1+q)u|_0-qu|_b=0, \quad u|_0-u|_b=0. \quad \square \tag{4}$$

For the operation $\mathbf{L}(D)=D_x^2$, $x\in(0,b)$, an analog of the first equation (4) is given by the equations

$$[p_1u+(1+q_1)u']_0-[p_1u+(bp_1+q_1)u']_b=0,$$
$$[(1+p_2)u+q_2u']_0-[p_2u+(bp_2+q_2)u']_b=0,$$

which, for arbitrary p_1, p_2, q_1, q_2, describe all proper operators (in the strict sense) generated by D_x^2. It is easy to write the corresponding conditions for $D_x^2-\lambda$ (analog of (2)). However, it would be tedious to describe, even in this simple case, the form of a "minimal" set of conditions (i.e., to write them in the most economical form) of the type of (4), to describe all operators that are proper in the wide sense and generated by D_x^2. In any case, it is clear from the preceding discussion that in principle their complete description requires four arbitrary parameters (contained, generally speaking, in a system of equations). The introduction of a fifth parameter λ would lead to the appearance of equivalent conditions.

We note that writing the boundary conditions in the form (Γ) of §3, even after normalization, introduces a significantly larger number of parameters.

Chapter IV
Model Operators

§ 0. Introductory Remarks

As we remarked in the preceding chapter, the main classes of objects that are of interest here are the equations obtained from equations for ordinary differential operations by replacing coefficients that are functions by coefficients that are operators. The following use of formulas that provide solutions of ordinary equations in order to clarify properties of the solutions of operator equations turns out to be unavoidably connected with the study of functions of operators, i.e. with the construction of an operational calculus.

The construction of this calculus assumed, in the first instance (2.2, Chapter I), our knowledge of the spectrum of the operator of which functions were to be defined. In the present chapter we consider a class of operators whose spectra can be completely described. These are the "model" operators which will be extensively used in what follows.

In § 1 we consider the general construction, and in § 2 we make a detailed study of the simplest model operators, the linear differential operators, with constant coefficients, on the n-dimensional torus. The method that we use to study these operators (Π-operators) is also the prototype of the constructions that will be used later on.

§ 1. Tensor Products and Model Operators

1.1. Tensor Products of Hilbert Spaces. Let $\mathbb{H}'$ and $\mathbb{H}''$ be two separable Hilbert spaces, in each of which we are given orthonormal bases $\{\varphi_k\}_1^\infty$, $\{\psi_k\}_1^\infty$. We construct a Hilbert space $\mathbb{H}$ in the following way. As a basis in $\mathbb{H}$ we take the set of ordered pairs $\varphi_k \otimes \psi_j$, defining the scalar product of pairs by the rule

$$(\varphi_k \otimes \psi_j, \, \varphi_l \otimes \psi_i) = (\varphi_k, \varphi_l)(\psi_j, \psi_i), \tag{1}$$

where on the right we have the scalar products in $\mathbb{H}'$ and $\mathbb{H}''$. Consequently, with the norm generated by the scalar product (1), the basis

$\{\varphi_k \otimes \psi_j\}_{k,j}$ is orthonormal. The product (1) can be extended in the usual way to linear combinations

$$\sum f_{kj}\, \varphi_k \otimes \psi_j. \tag{2}$$

The completion under the norm of the set of finite linear combinations (2) yields a (complete) Hilbert space $\mathbb{H} = \mathbb{H}' \otimes \mathbb{H}''$, the *tensor product* of the original *Hilbert spaces*.

In agreement with the preceding construction we can define, for any two elements $f = \sum f_k \varphi_k \in \mathbb{H}'$ and $g = \sum g_k \psi_k \in \mathbb{H}''$, their tensor product

$$f \otimes g = \sum_{i,k} f_i\, g_k\, \varphi_i \otimes \psi_k$$

(since $\sum_{i,k} |f_i|^2\, |g_k|^2 < \infty$).

If now $\mathbf{A}'\colon \mathbb{H}' \to \mathbb{H}'$ is a closed linear operator with dense domain $\mathfrak{D}(\mathbf{A}')$, $\varphi_k \in \mathfrak{D}(\mathbf{A}')$, for every k, and the operator $\mathbf{A}''\colon \mathbb{H}'' \to \mathbb{H}''$ has the corresponding properties, then we can define, on a set of elements of the form (2) that is dense in $\mathbb{H}$ (on the set of finite linear combinations), the operator

$$\mathbf{A}' \otimes \mathbf{A}'' \left(\sum f_{ik}\, \varphi_i \otimes \psi_k \right) = \sum f_{ik}\, \mathbf{A}'\, \varphi_i \otimes \mathbf{A}''\, \psi_k.$$

The closure in $\mathbb{H}$ of the operator $\mathbf{A}' \otimes \mathbf{A}''$ defined in this way (with dense domain) defines the operator $\mathbf{A}' \otimes \mathbf{A}''\colon \mathbb{H} \to \mathbb{H}$.

If $\mathbb{H} = \mathbb{H}' \otimes \mathbb{H}''$ and $\mathbb{H}'$, $\mathbb{H}''$ are function spaces, then $\mathbb{H}'$ can be embedded in $\mathbb{H}$ in a natural way, by identification with the subset $\mathbb{H}' \otimes 1$ (composed of the elements of the form $f \otimes 1$, $f \in \mathbb{H}'$). With this in mind, the elements of $\mathbb{H}'$ are often thought of as elements of $\mathbb{H}$ without any reservations (and without change of notation from f to $f \otimes 1$). Similarly, the operators $\mathbf{A}'\colon \mathbb{H}' \to \mathbb{H}'$ are identified with operators of the form $\mathbf{A}' \otimes 1$.

These constructions arise in a natural way whenever $\mathbb{H}'$ and $\mathbb{H}''$ are our standard Hilbert spaces over domains $V' \subset \mathbb{R}^{n'}$ and $V'' \subset \mathbb{R}^{n''}$. Here $\mathbb{H}$ is the corresponding space over $V' \times V''$. Then the operation $\mathbf{L}'(D) \otimes \mathbf{L}''(D)$ and the corresponding operator are usually simply written in the form

$$\sum a_\alpha(x)\, b_\beta(y)\, D_x^\alpha D_y^\beta,$$

i.e. without the use of the symbol $\otimes$ for the tensor product.

Since in a Hilbert space the transition from a Riesz basis to an orthonormal basis, and conversely, is equivalent to the change from a given scalar product to an equivalent one (see [1]), it is clear that the preceding discussion can be extended in a natural way to the case when $\{\varphi_k\}$ and $\{\psi_j\}$ are Riesz bases in $\mathbb{H}'$ and $\mathbb{H}''$.

The transition from $\mathbb{H} = \mathbb{H}' \otimes \mathbb{H}''$ to an arbitrary number of factors, $\mathbb{H} = \bigotimes_{k=1}^{n} \mathbb{H}^k$, is trivial.

1.2. Model Operators. The M-operators (3.5, Chapter I) form a convenient class of operators, functions of which have a very simple definition. In fact, if $\mathbf{A}: \mathbb{H} \to \mathbb{H}$ is an M-operator, and $\{\varphi_k\}$ is a system of eigenfunctions of $\mathbf{A}$ that form a Riesz basis, and consequently we can write, for each $u \in D(A)$,

$$u = \sum u_k \varphi_k, \qquad \mathbf{A} u = \sum u_k \mathbf{A} \varphi_k = \sum \lambda_k u_k \varphi_k;$$

then, under the assumption that, for example, $F(z)$ is analytic in a domain $\Omega \subset \mathbb{C}$ such that $\lambda_k \in \Omega$ for all k, we have only to set

$$F(\mathbf{A}) u = \sum F(\lambda_k) u_k \varphi_k. \tag{3}$$

Here $u \in \mathfrak{D}(F_A)$ whenever the series (3) converges. The domain of $F(\mathbf{A})$ is necessarily dense (it contains all finite linear combinations of the elements of the basis), and our definition is consistent with the construction in 2.2, Chapter I, and with the ordinary connotations of "operational calculus".

The difficulties that are encountered in trying to apply this idealized scheme to specific situations that arise in the study of boundary value problems are ordinarily connected, on the one hand, with the complicated nature of $F(z)$, and on the other hand with the tendency toward considering operators that are not M-operators (see Chapter VIII).

The preceding scheme extends immediately to the case when $\mathbb{H} = \overset{n}{\underset{1}{\bigotimes}} \mathbb{H}_k$, $\mathbf{A}^k: \mathbb{H}^k \to \mathbb{H}^k$ and $F(z_1, \ldots, z_n)$ is a function of n complex variables that satisfies corresponding requirements. The operators $\mathbf{A}^k$ are naturally understood to be commutative.

§2. Operators on the n-Dimensional Torus

2.1. Definition of Π-Operators and their Basic Properties. The simplest "explicit" example of the situation described above consists of operators on the n-dimensional torus, generated by differential operations of the form (1), §2 of Chapter II, with constant coefficients. From a different point of view, these are operators with constant coefficients considered in a parallelepiped $V \subset \mathbb{R}^n$ and acting on functions that are periodic in all variables. Let us describe these operators in more detail.

It is convenient to suppose that V is simply the n-dimensional cube with edges of length 2π (some remarks on the possible effects of a change in the parameters of V will be made later). Let $\mathscr{P}^\infty$ be the linear manifold of smooth complex-valued functions that are periodic in each variable, and let $\mathbb{H}(V)$ be our usual Hilbert space in which $\mathscr{P}^\infty$ is dense. With a polynomial $A(s)$ with constant coefficients,

$$A(s) = \sum_{|\alpha| \leq m} a_\alpha s^\alpha, \qquad s^\alpha = s_1^{\alpha_1} \ldots s_n^{\alpha_n}, \qquad |\alpha| = \alpha_1 + \ldots + \alpha_n, \tag{1}$$

we associate a differential operation $A(-iD)$ so that

$$A(-iD)\,e^{is\cdot x} = A(s)\,e^{is\cdot x}, \qquad s\cdot x = s_1 x_1 + \ldots + s_n x_n.$$

We define the corresponding operator $A: \mathbb{H} \to \mathbb{H}$ to be the closure in $\mathbb{H}$ of the operation $A(-iD)$, defined in the first instance on functions that belong to $\mathscr{P}^\infty$. Operators A with the structure described above are called Π-operators.

Remark. By definition, the n-dimensional torus $\mathbb{T}^n$ is the direct product of n circles $\mathbb{T}^1$, and our Hilbert space $\mathbb{H}(V)$ automatically has the structure of the tensor product of n spaces $\mathbb{H}^1$ over $\mathbb{T}^1$. A corresponding remark can be made about our Π-operators A, which are representable as sums of monomials each of which is the tensor product of powers of the operator D_x considered on a circle. This also determines the specific structure of the operators in $\mathbb{H}$. Subsequent constructions, however, will be based on the "direct" study of operators A without introducing the terminology of tensor products.

It will be convenient to denote by $\mathscr{S}$ the set of n-dimensional integral-valued vectors $(s_1, \ldots, s_n)$, $s_k = 0, \pm 1, \pm 2, \ldots$. The set of exponentials $\{e^{is\cdot x}\}$, $s \in \mathscr{S}$, $s\cdot x = s_1 x_1 + \ldots + s_n x_n$, evidently forms an orthogonal basis in $\mathbb{H}$ and is at the same time a set of eigenelements for each operator A. For a given operator A (generated by the operation $A(-iD)$) each number $A(s)$, $s \in \mathscr{S}$, is an eigenvalue. We denote the set of these numbers by $A(\mathscr{S})$.

Proposition 1. *Each pair* A_1, A_2 *of* Π-operators commute.

This proposition follows immediately from the commutativity of the operations $A_k(-iD)$.

Proposition 2. *Every* Π-operator A is normal. (3.4, Chapter I).

Proof. If $A^t: \mathbb{H} \to \mathbb{H}$ is an operator generated under our hypotheses by the operation $A^t(-iD)$, the equation $A^t A = A A^t$ follows from Proposition 1. It is therefore enough to show that $A^* = A^t$, i.e. that the weak and strong definitions of A^t are equivalent. However, for an operation with constant coefficients, considered on smooth periodic functions, $A(-iD)J_\varepsilon u = J_\varepsilon A(-iD)u$ (where J_ε is a standard averaging operator) and the requirement of equivalence follows immediately from §6 of Chapter I.   □

Proposition 2 can evidently also be obtained from the representation of Au in the form

$$A u = \sum_s A(s)\,u_s\,e^{is\cdot x}, \tag{2}$$

where

$$u(x) = \sum_s u_s\,e^{is\cdot x} \in \mathfrak{D}(A) \tag{3}$$

(we have used the general scheme of 1.2).

The representation (2) immediately lets us pass to the consideration of operators $F(\mathbf{A})$ where $F(z)$ is a function that takes finite values on $\mathbf{A}(\mathscr{S})$. A corresponding operator is defined by the equation

$$F(\mathbf{A})\,u = \sum_s F(\mathbf{A}_s)\,u_s\,e^{is\cdot x}, \qquad (4)$$

where $\mathbf{A}_s = \mathbf{A}(s)$. As in 1.2, $u\in\mathfrak{D}(F)$ if the series (4) converges in $\mathbb{H}$.

We denote the operator (4) simply by F, and the set $F(\mathbf{A}_s)$, $s\in\mathscr{S}$, by $F(\mathscr{S})$.

Proposition 3. *The spectrum σF of the operator $F: \mathbb{H}\to\mathbb{H}$ is the closure in the complex plane of the set $F(\mathscr{S})$ formed by the point spectrum $P\sigma F$ of F. The set $C\sigma F = \sigma F\backslash P\sigma F$ is the continuous spectrum of F.*

Proof. Denote $F(\mathbf{A}_s)$ by $F(s)$. If $\lambda = F(s)$ for some $s\in\mathscr{S}$, then $(F-\lambda)\,e^{is\cdot x} = 0$, i.e. λ is an eigenvalue of F. If $\lambda\notin F(\mathscr{S})$ the operator F_λ^{-1} exists. In fact, in this case $F_\lambda u = 0$ implies $u_s = 0$ for every $s\in\mathscr{S}$, i.e. $u = 0$ (uniqueness of the Fourier expansion). Moreover, in this case the set $\mathfrak{D}(F_\lambda^{-1})$ is dense in $\mathbb{H}$, since it contains all finite sums of the form (3).

Here, if the condition $|\lambda - F(s)|\geq\delta>0$ is satisfied for every $s\in\mathscr{S}$ (i.e., $\lambda\notin\overline{F(\mathscr{S})}$), the operator F_λ^{-1} is bounded and defined on $\mathbb{H}$ by

$$(F-\lambda)^{-1}f = \sum_s (F(s)-\lambda)^{-1} f_s\, e^{is\cdot x}$$

(it is assumed that f is represented by an expansion of the form (3)). If $\lambda\in\overline{F(\mathscr{S})}\backslash F(\mathscr{S})$, there exists a sequence $\{s_k\}$ such that $|F(s_k)-\lambda| = \varepsilon_k\to 0$ as $k\to\infty$ and the corresponding operator F_λ^{-1} will be unbounded (since $\|F_\lambda^{-1}\,e^{is_k\cdot x}\|/\|e^{is_k\cdot x}\| = \varepsilon_k^{-1}$). $\square$

Therefore the structure of the spectrum of $F(\mathbf{A})$ is quite transparent. Unfortunately, the study of the corresponding questions is significantly more complicated for the families of operators $F(\mathbf{A},t)$, $t\in(0,b)$, which are encountered in discussing operator equations.

The simplicity of the operators $F(\mathbf{A})$, and of the operator $\mathbf{A}$ itself $(F(z)\equiv z)$ does not prevent them, in many cases, from being "bad", i.e. from having an empty resolvent set.

Proposition 4. *For $n\leq 2$ the set $\rho\mathbf{A}$ for a Π-operator $\mathbf{A}$ is always nonempty; for $n>2$ there exist operators $\mathbf{A}$ whose spectra fill the entire complex plane.*

Proof. For $n=1$ the conclusion is trivial; for $n=2$ it is enough to consider the mapping from the real (s_1, s_2) plane on the complex plane of $z = x+iy$ defined by the equations

$$x = \operatorname{Re}\mathbf{A}(s_1, s_2), \qquad y = \operatorname{Im}\mathbf{A}(s_1, s_2).$$

Because of the algebraic nature of the mapping, the image of the set of (s_1, s_2) points with integral coordinates cannot be dense in the z plane.

When $n = 3$ the set of values of the polynomial

$$A(s) = s_1 + \alpha s_2 + i(s_3 + \beta s_2^2),$$

where α and β are irrational, is dense in the complex plane. This follows from the uniform distribution in the unit square of the fractional parts of the pair $(\alpha s, \beta s^2)$ for $s = 0, \pm 1, \pm 2, \ldots$ (see [4])[2]. $\square$

For $s \in \mathscr{S}$, we define $|s|$ in the usual way by $|s|^2 = \sum s_k^2$.

Proposition 5. *The operator* A^{-1}: $\mathbb{H} \to \mathbb{H}$ *inverse to a* Π-*operator* A *is a CC operator* (3.1, Chapter I) *if and only if* $|A(s)| \to \infty$ *when* $|s| \to \infty$.

Proof. Necessity. If the stated condition on A is not satisfied, then for some $M > 0$ there exist an infinite number of values $s \in \mathscr{S}$ such that $|A(s)| \leq M$. If $\lambda_s = A(s)$ is an infinite set of corresponding eigenvalues, then $\mu_s = \lambda_s^{-1}$ are eigenvalues of A^{-1} ($\lambda_s \neq 0$ since, by hypothesis, A^{-1} exists). We have $|\mu_s| > M^{-1}$, which contradicts the complete continuity of A^{-1} (Lemma 1, 3.2 of Chapter I).

Sufficiency. Let the condition of the conclusion be satisfied. Then for each $\varepsilon > 0$, if we choose $M(\varepsilon)$ so that $(2\pi)^n M^{-2} < \varepsilon^2$, and $N(M)$ so that $|A(s)| \geq M$ for $|s| \geq N(M)$, we will have

$$A^{-1} = A_N^{-1} + A_R^{-1}, \tag{5}$$

where $A_N^{-1} f = \sum\limits_{|s| < N} (f_s/A(s)) e^{is \cdot x}$ is a finite-dimensional operator, and the norm of the second term in (5) (the remainder of the series) satisfies the inequality

$$\|A_R^{-1} f\|^2 = \Big\| \sum\limits_{|s| \geq N} (f_s/A(s)) e^{is \cdot x} \Big\|^2 \leq \frac{(2\pi)^n}{M^2} \sum\limits_{|s| \geq N} |f_s|^2 \leq \frac{(2\pi)^n}{M^2} \|f\|^2 < \varepsilon^2 \|f\|^2.$$

It follows from this and the approximation theorem (3.1, Chapter I) that A^{-1} is completely continuous. $\square$

Proposition 5 can evidently be stated differently, as follows.

Proposition 5'. *The inverse* A^{-1}: $\mathbb{H} \to \mathbb{H}$ *of a* Π-*operator* A *is a CC operator if and only if, for every* $N > 0$ *there is only a finite set of linearly independent eigenfunctions of* A *that belong to the eigenvalues* λ_s *that satisfy* $|\lambda_s| \leq N$. $\square$

---

[2] This number-theoretic proposition was called to the author's attention by Professor A.G. Postnikov.

Therefore for our class of operators $\mathbf{A}^{-1}$, Lemma 1 of 3.2, Chapter I, provides a necessary and sufficient condition for complete continuity. Consequently the failure of complete continuity for $\mathbf{A}^{-1}$ must imply either that $\mathbf{A}$ has infinitely many eigenvalues in a finite domain, or that the eigenspace belonging to some λ_s is infinite-dimensional. As we shall see in the next subsection, the complete continuity of $\mathbf{A}^{-1}$ is closely connected with the "stability" of $\mathbf{A}$.

The following proposition describes the differential properties of elements that are solutions of equations containing Π-operators.

Proposition 6. *Let* $\mathbf{A}$ *and* $\mathbf{B}$ *be* Π*-operators and let* $\mathbf{A}^{-1}$ *exist. Then a necessary and sufficient condition for* $\mathbf{B}\mathbf{A}^{-1}$ *to be a bounded operator that admits extension to all of* $\mathbb{H}$ *is that*

$$|\mathbf{B}(s)/\mathbf{A}(s)| \leq M < \infty \quad \text{for all } s \in \mathscr{S}. \tag{6}$$

Proof. The condition is evidently sufficient since it guarantees that $\mathbf{B}$ can be applied termwise to the series that represents the element $u = \mathbf{A}^{-1}f$, as well as the boundedness of $\mathbf{B}\mathbf{A}^{-1}$. Since the domain of $\mathbf{A}^{-1}$ is necessarily dense in $\mathbb{H}$, the operator $\mathbf{B}\mathbf{A}^{-1}$ can be extended to the whole space.

The necessity of (6) follows because the elements of the basis $\{e^{is \cdot x}\}$, $s \in \mathscr{S}$, are eigenfunctions of $\mathbf{B}\mathbf{A}^{-1}$, and the numbers $\mathbf{B}(s)/\mathbf{A}(s)$ are the corresponding eigenvalues. An operator cannot be bounded when there is an unboundedly increasing sequence of eigenvalues. $\square$

Proposition 6 can evidently be reformulated in terms of having the solution $u = \mathbf{A}^{-1}f$ belong to some space of type W (§7, Chapter II).

It is appropriate also to mention the following particular property. Although when $\mathbf{B}(s) \neq \text{const.}$ the operator $\mathbf{B}$ is necessarily unbounded, $\mathbf{B}\mathbf{A}^{-1}$ can be bounded even when $\mathbf{A}^{-1}$ is unbounded. Trivial case: $\mathbf{B} = \mathbf{A}$. A somewhat less trivial case: $\mathbf{A} = \mathbf{A}_1 \mathbf{A}_2$, where $\mathbf{A}_1^{-1}$ is bounded and the unboundedness of $\mathbf{A}^{-1}$ is due to the unboundedness of $\mathbf{A}_2^{-1}$ (for example, $\mathbf{A}_2(s) = s_1 + \alpha s_2$, where α is irrational). Then $\mathbf{B}\mathbf{A}^{-1}$ is bounded when $\mathbf{B} = \mathbf{A}_2$.

2.2. Some Further Properties of Π-Operators. We continue to consider Π-operators, and establish further properties that will be useful in what follows.

In many cases it is convenient to consider $\mathbf{A}(s)$ as a sum

$$\mathbf{A}(s) = \mathbf{R}(s) + i \mathbf{Q}(s) \tag{7}$$

where $\mathbf{R}$ and $\mathbf{Q}$ are polynomials with real coefficients. This decomposition of $\mathbf{A}(s)$ evidently corresponds to the decomposition of a normal operator $\mathbf{A}$: $\mathbb{H}_x \rightarrow \mathbb{H}_x$ into symmetric and skew-symmetric parts.

Preserving the notation of the preceding subsection, we consider, for real polynomials $\mathbf{R}(s)$, the relations

$$\lim_{s\in\mathscr{S},\;|s|\to\infty} |\mathbf{R}(s)|=\infty, \tag{C}$$

$$\inf_{s\in\mathscr{S}}\mathbf{R}(s)\geq -M> -\infty, \qquad \sup_{s\in\mathscr{S}}\mathbf{R}(s)\leq M< +\infty. \tag{B}$$

We say that $\mathbf{R}$ has the C-property (or B-property) if it satisfies (C) (or one of the relations (B)).

For $n=1$, every polynomial has the C-property, and has the B-property if and only if its leading term has even degree. For $n\geq 2$, not every polynomial has the C-property. The nature of the "compact" case, which is connected with the use of Fourier series (but not of Fourier integrals) and a discrete set $\mathscr{S}$, is reflected in the following proposition.

Proposition 7. *For $n\geq 2$ the preceding properties of the polynomial $\mathbf{R}(s)$ are independent.*

Proof. That (B) does not, in general, imply (C) is rather obvious (for example, $\mathbf{R}(s)=(s_1-s_2)^2$). For an example of a polynomial that has the C-property but not the B-property, we may use $\mathbf{R}(s)=(s_1+\frac{1}{2})(s_2+\frac{1}{2})$. This evidently does not have the B-property. However, since $|s_k+\frac{1}{2}|\geq |s_k|/2$, and at the same time $|s_k+\frac{1}{2}|\geq\frac{1}{2}$, $s_k=0,\pm 1,\pm 2,\ldots$, the C-property follows from the inequality $|\mathbf{R}(s)|\geq |s_k|/4$, which implies that $|\mathbf{R}(s)|\geq (|s_1|+|s_2|)/8$. $\square$

Remark. If we waive the condition $s\in\mathscr{S}$ in (C), and use arbitrary values $s\in\mathbb{R}^n$, then (C) implies (B) for $n\geq 2$, since in this case the C-property is equivalent to the existence of the limit $\lim_{|s|\to\infty}\mathbf{R}(s)=\pm\infty$. This follows because a real polynomial that takes values of opposite sign outside a ball also takes the value zero outside this ball, since when $n>1$ the complement of a ball is connected.

The possession of the B-property by one of $\mathbf{R}$ or $\mathbf{Q}$ in (7) immediately guarantees the existence of a half plane that is free of points of the spectrum of $\mathbf{A}$. However, as we shall now see, this does not guarantee any (even very weak) property of "stability" of the spectrum under perturbations of the operator. Let us introduce the definition of stability.

Definition. An operator $\mathbf{A}$ is *stable* with respect to the operator $\mathbf{A}_0$ if, for every $z\in\rho\mathbf{A}$, there exists $\delta=\delta(\mathbf{A}_0,z)>0$ such that the condition $|\varepsilon|<\delta$ implies that $z\in\rho(\mathbf{A}+\varepsilon\mathbf{A}_0)$.

Proposition 8. *If there exist $\gamma>0$ and $M=M(\gamma)$ such that the requirement*

$$|s|\geq M \quad \text{implies} \quad |\mathbf{A}(s)/\mathbf{A}_0(s)|\geq\gamma \tag{8}$$

is satisfied for every $s\in\mathscr{S}$, then $\mathbf{A}$ is stable with respect to $\mathbf{A}_0$.

Proof. Let $z\in\rho\mathbf{A}$ and consequently let there exist $r>0$ such that

$$|z-\mathbf{A}(s)|\geq r>0$$

for every $s\in\mathscr{S}$.

Assuming that (8) is satisfied, we first find conditions on ε under which $z \notin P\sigma(A + \varepsilon A_0)$. Suppose that $z \in P\sigma(A + \varepsilon A_0)$, i.e. that for some $s \in \mathscr{S}$ the equation $A(s) + \varepsilon A_0(s) = z$ is satisfied. This is possible only when $A_0(s) \neq 0$, and then we can write

$$|\varepsilon| = \frac{|z - A(s)|}{|A_0(s)|} \geq \frac{r}{|A_0(s)|} \geq \frac{\gamma}{2}. \tag{9}$$

Now choose a number N so large that, first,

$$|A_0(s)| < N \tag{10}$$

for $|s| < M$, and, second,

$$\gamma - \frac{|z|}{N} \geq \frac{\gamma}{2}. \tag{11}$$

Now if (10) is satisfied it follows from (9) that

$$|\varepsilon| \geq r N^{-1}$$

for $|s| < M$. If, however, $|A_0(s)| \geq N$, and consequently $|s| \geq M$, then, by using (8), (9) and (11) we obtain

$$|\varepsilon| = \frac{|z - A(s)|}{|A_0(s)|} \geq \frac{|A(s)|}{|A_0(s)|} - \frac{|z|}{N} \geq \frac{\gamma}{2}.$$

Consequently when $|\varepsilon| < \delta = \min(r/N, \gamma/2)$ the point z cannot belong to $P\sigma(A + \varepsilon A_0)$.

If we now suppose that $z \in C\sigma(A + \varepsilon A_0)$, there must be a sequence $\{s^k\} \in \mathscr{S}$ such that

$$A(s^k) + \varepsilon A_0(s^k) = z + \eta_k, \quad |\eta_k| \to 0 \quad \text{as } k \to \infty.$$

Then for sufficiently large indices k we must have

$$|z + \eta_k - A(s^k)| \geq \frac{r}{2} > 0, \quad |z + \eta_k| \leq |z| + 1,$$

and, repeating the preceding argument, we can find $\delta > 0$ such that the condition $|\varepsilon| < \delta$ implies $z \in \rho(A + \varepsilon A_0)$. $\quad\square$

Proposition 9. *If there is a sequence $\{s^k\} \in \mathscr{S}$ such that*

$$|A(s^k)/A_0(s^k)| = \delta_k \to 0, \quad |A_0(s_k)| \to \infty \quad \text{as } k \to \infty,$$

then A is unstable with respect to A_0.

In fact, if the preceding conditions are satisfied for every $z \in \rho \mathbf{A}$ we can evidently satisfy the equation

$$\mathbf{A}(s) + \varepsilon \mathbf{A}_0(s) = z$$

for arbitrarily small ε. $\square$

It is clear from (9) that, for example, for $\mathbf{R}(s)$ to have the C-property is, in some sense, necessary for the corresponding operator $\mathbf{R}$ to have at least a minimal stability property. More precisely, we have the following proposition.

Proposition 10. *If* $\mathbf{R}(s)$ *does not have the C-property, then the operator* $\mathbf{R}$ *is unstable, generally speaking, with respect to every operator* $\mathbf{R}_0$ *of first order.*

Proof. By hypothesis there is an infinite sequence $\{s^k\} \in \mathscr{S}$ such that $|s^k| \to \infty$ as $k \to \infty$, but $|\mathbf{R}(s^k)| \leq M < \infty$. Let $\{\varphi^k\}$ be the set of intersections of the rays Os^k (from the origin to the points s^k) with the unit sphere. Then, as follows immediately from (9), the operator $\mathbf{R}$ is unstable with respect to every operator $\mathbf{R}_0$, $\mathbf{R}_0(s) = \sum a^l s_l$, so that there is a limit point φ of the sequence $\{\varphi^k\}$ that does not lie on the intersection of the hyperplane $\sum a^l s_l = 0$ with the unit sphere. $\square$

Another interpretation of Proposition 9 is the necessity of requiring the complete continuity of $\mathbf{A}^{-1}$ under reasonable stability properties of the Π-operator $\mathbf{A}$.

Remark. In considering questions connected with stability, it is natural to include the study of the possibility of the influence of small perturbations of the cube $[0, 2\pi]^n$ on the solvability of the problem. Evidently the passage from V to, for example, the parallelepiped $V_\varepsilon = [0, 2\pi]^{n-1} \times [0 \leq x_n \leq 2\pi - \varepsilon]$ (preserving periodicity) is equivalent to applying the factor $((2\pi - \varepsilon)/2\pi)^k$ to the coefficients of the monomials in $\mathbf{A}(s)$ that contain the factor $(s_n)^k$. Similar perturbations are included, therefore, in the line of investigation discussed above.

In conclusion, we note that in considering the dependence of the spectrum of $\mathbf{A}$ on properties of $\mathbf{R}(s)$ and $\mathbf{Q}(s)$, the following property plays a fundamental role. Let $\mathbf{R}(s)$ depend on a set of variables $s' \in \mathscr{S}' \subset \mathscr{S}$, and let $\mathbf{Q}(s)$ depend on a set of variables $s'' \in \mathscr{S}'' \subset \mathscr{S}$ (for example, when $n=3$, let $\mathbf{R}(s) = s_1 + s_2^2$ and $\mathbf{Q}(s) = s_1 + s_3$). We say that

    a) $\mathbf{Q}(s)$ depends completely on $\mathbf{R}(s)$,
    b) $\mathbf{Q}(s)$ depends partially on $\mathbf{R}(s)$,
    c) $\mathbf{Q}(s)$ is independent of $\mathbf{R}(s)$,

if

    a) the sets $\mathscr{S}'$ and $\mathscr{S}''$ are connected by the relation $\mathscr{S}' \supset \mathscr{S}''$,
    b) the sets $\mathscr{S}'$ and $\mathscr{S}''$ have nonempty intersection, but the inclusion $\mathscr{S}'' \subset \mathscr{S}'$ does not hold,
    c) the intersection $\mathscr{S}' \cap \mathscr{S}''$ is empty.

When $n=3$ the polynomials $\mathbf{R}(s)=s_1+s_2^2$ and $\mathbf{Q}(s)=s_1+s_2$; s_1+s_3; s_3 illustrate the cases a), b) and c).

If the skew-symmetric part $\mathbf{Q}$ of the operator $\mathbf{A}$ is completely dependent on $\mathbf{R}$, the spectrum of $\mathbf{A}$ lies completely on a regular curve $y=\varphi(x)$ in the complex plane of $z=x+iy$. However, as was shown in 2.1, in the case of partial dependence of $\mathbf{Q}(s)$ the set $\rho\mathbf{A}$ can reduce to the empty set when $n\geq 3$.

2.3. Π-Operators Generated by Some Classical Differential Operations. In the light of the preceding discussions it is instructive to notice the properties of the Π-operators generated by some differential operators of classical types.

Thus, for example, it is clear that a special place is occupied by the operators $\mathbf{A}(-iD)$ of order $2m$ with real coefficients whose principal homogeneous parts are defined by polynomials with the property that

$$\sum_{|\alpha|=2m} a_\alpha s^\alpha \geq c \sum_{k=1} s_k^{2m}, \quad c>0.$$

A corresponding operator $\mathbf{A}^{-1}$ is completely continuous, with a pure point spectrum that is sufficiently sparse and stable with respect to perturbations by any operator $\mathbf{A}_0$ of order $\leq 2m$. From the classical point of view we are dealing here with a subclass of the elliptic operators. Similar properties are possessed by the quasielliptic operators, for which the coefficients of the corresponding polynomials are real and satisfy the equations

$$\sum_{|\alpha|\leq 2m} a_\alpha s^\alpha \geq c \sum_{k=1}^{n} s_k^{2m_k}, \quad m_k\geq 1.$$

An example of an elliptic nonpositive polynomial is $\mathbf{A}(s)=s_1+is_2$. The corresponding (Cauchy-Riemann) operator also has very nice properties, but this situation is peculiar to dimension $n=2$.

We may say that $\mathbf{A}(s)$ has the strong C-property if

$$\lim_{s\in\mathbb{R}^n,\ |s|\to\infty} |\mathbf{A}(s)| = \infty.$$

The corresponding operator belongs to the class of hypoelliptic operators. A typical example is the polynomial $is_1\pm s_2^2$ corresponding to the direct and inverse operators of heat conduction. The polynomial $\mathbf{A}(s)$ is essentially inhomogeneous and contains both a real and an imaginary component.

The nice properties of our classes of Π-operators (complete continuity of $\mathbf{A}^{-1}$, stability of the spectrum) are closely connected with the reasonableness for these classes of conditions like periodicity in all variables. The picture changes substantially for Π-operators corresponding to classical operations of hyperbolic type.

Typical representatives of the latter classes of polynomials are

$$s_1 + \sum_2^n a^k s_k, \qquad s_1^2 - \sum_2^n s_k^2 \tag{12}$$

(a^k real). The Π-operators generated by these polynomials are "bad" (A^{-1} not completely continuous; unstable spectrum). For operations corresponding to the polynomials (12) the "nice" operators are those defined by conditions of periodicity in $s_2, \ldots, s_n$ and qualitatively different conditions (Cauchy conditions, which are irregular from the point of view of §3, Chapter III) on a selected variable, corresponding to s_1, and playing the role of time. We shall take up the corresponding constructions in the next chapter.

At the same time, we notice that to obtain proper operators generated, for example, by the operation

$$\sum_1^n (a^k + i c^k) D_k$$

with real a^k and c^k, $n \geq 5$, we have, generally speaking, to call on boundary conditions of a quite unusual nature. We shall discuss these in §2 of Chapter V.

Chapter V
First-Order Operator Equations

§0. Introductory Remarks

This chapter is, in a definite sense, central to our entire exposition. Here, by using the simplest entities, operator equations of the first order, generated by special classes of boundary value problems for partial differential equations that originate from the constructions and results presented above, we consider the whole circle of ideas that are of interest. In passing to more complicated entities the study of the corresponding questions turns out to involve a whole series of difficulties of a technical nature, whereas the constructions in this chapter are as transparent and elementary as possible.

Let $t \in (0, b) = V_t$, $\mathbb{H}_t \equiv \mathbb{H}_t(V_t)$ the corresponding Hilbert space, and D_t: $\mathbb{H}_t \to \mathbb{H}_t$, any of the operators generated by the operation D_t. Then by an operator equation (or differential-operator equation) of the first order we ordinarily understand an equation of the form

$$\mathbf{L} u \equiv (\mathbf{A}_0 D_t + \mathbf{A}_1) u = f, \tag{1}$$

$\mathbf{A}_0, \mathbf{A}_1: \mathcal{B} \to \mathcal{B}$, where $\mathcal{B}$ is a B-space and $\mathbf{A}_k$ are operators, commuting with D_t and with domains dense in $\mathcal{B}$. Here the operator $\mathbf{L}$ is considered to act on the space $\mathbb{H} = \mathbb{H}_t \otimes \mathcal{B}$.

Since our fundamental objects of study are boundary value problems, the role of $\mathcal{B}$ is taken in most cases by the space $\mathbb{H}_x \equiv \mathbb{H}_x(V)$, where V is a bounded domain in the space of $x \in \mathbb{R}^n$. The operators $\mathbf{A}_k$ are the operators generated by the operations $\mathbf{A}_k(-iD)$ on V. Moreover, the main part of the present chapter concerns the case when $\mathbf{A}_k$ is a Π-operator (Chapter IV), and the operator $\mathbf{A}_0 \equiv 1$. The possibility of transferring the results to the case when $\mathbf{A}_0$ and $\mathbf{A}_1$ are arbitrary M-operators (3.5, Chapter I) whose spectra satisfy the corresponding requirements is rather evident and will not be discussed separately. Chapter VIII is devoted to the discussion of operator equations under significantly weaker hypotheses on the operators (stated in terms of properties of resolvents).

With the specializations of D_t and $\mathbf{A}_k$ indicated above, the operator $\mathbf{L}$ in (1) turns out to be generated by certain differential operations $\mathbf{L}(D)$ on $V_t \times V$. In the course of the chapter we will see numerous examples of differential operators along with an investigation of their properties. These examples are models in a sense that corresponds to the study of the classical equations of mathematical physics under assumptions that simplify this study as much as possible. However, our approach allows us to include at the same time a wide class of nonclassical operations and to clarify a number of special properties of the solutions of boundary value problems.

The first three sections are devoted to a detailed study of the simplest operator equations, and the rest, to various special questions.

§1. The Operator $D_t - \mathbf{A}$; the Spectrum

As we noted in the introduction, the simplest operator equation of the first order is of the form

$$\mathbf{L} u \equiv (D_t - \mathbf{A}) u = f, \tag{L}$$

where $f \in \mathbb{H} = \mathbb{H}_t \otimes \mathbb{H}_x$, $D_t : \mathbb{H}_t \to \mathbb{H}_t$ is the regular operator generated by the condition

$$\mu u|_{t=0} - u|_{t=b} = 0 \tag{Γ}$$

for $t \in (0, b)$, and $\mathbf{A} : \mathbb{H}_x \to \mathbb{H}_x$ is a Π-operator (§2, Chapter IV). In accordance with the definitions of the operators that appear in (L), an element $u \in \mathbb{H}$ is called a solution of the $L - \Gamma$ problem if there is a sequence of smooth functions $u_i(x, t)$, converging to u in $\mathbb{H}$, such that u_i has period 2π in the variables x_k, satisfies condition (Γ) in t, and $\mathbf{L}(D) u_i = f_i \to f$ (convergence in $\mathbb{H}$) as $i \to \infty$.

With our hypotheses, the analysis of the $L - \Gamma$ problem for the operator $\mathbf{L}$ reduces in essence to the analysis of the operator defined by formula (11) of §2, Chapter III, with the numerical parameter λ replaced by the operator $\mathbf{A}$. However, the conclusions that we need cannot be obtained automatically from reasoning of the type used in 2.1, Chapter IV. Although the results are propositions that state that the properties of $\mathbf{L}$ are exhaustively described by properties of $\mathbf{A}$ (or of $\exp(b\mathbf{A})$), the proofs require a number of auxiliary constructions.

If we introduce the notation

$$\mathbf{T}(\lambda) = \exp b(\mathbf{A} + \lambda)$$

(the operator $\mathbf{T}(\lambda) : \mathbb{H}_x \to \mathbb{H}_x$ is to be thought of in the sense of the operations in §2, Chapter IV), we have the following theorem.

Theorem. *A point λ of the complex plane belongs to one of the sets $\rho \mathbf{L}$, $P\sigma \mathbf{L}$, $C\sigma \mathbf{L}$ if and only if the number μ in (Γ) belongs to $\rho \mathbf{T}(\lambda)$, $P\sigma \mathbf{T}(\lambda)$, $C\sigma \mathbf{T}(\lambda)$, respectively.*

Remark. For $\mu \neq 0$, ∞ the operator generated by D_t is proper, and the spectrum of L can be obtained from the theorem on the spectrum of a sum of commuting operators (see [6]). However, it is very important for us to include in the general picture the classical Cauchy problem ("inverse" for $\mu = 0$ and "direct" for $\mu = \infty$), which corresponds to the "irregular" case.

It follows from the hypotheses of the theorem that L, like A, has no residual spectrum.

The analysis of the $L - \Gamma$ problem, i.e. the proof of the theorem stated above, is based on reasoning similar to that used in §2, Chapter IV, in the study of the Π-operator A. If the spectrum of the Π-operator A is determined by the properties of the infinite chain of equations $(A(s) - \lambda) u_s = f_s$, $s \in \mathcal{S}$, then in the $L - \Gamma$ problem a similar role is played by the chain of ordinary differential equations

$$D_t u_s - A(s) u_s = f_s, \qquad s \in \mathcal{S} \tag{1}$$

(we preserve the notation of Chapter IV), where $u_s = u_s(t)$, and $f_s(t)$ are the coefficients (depending on t) of the expansions

$$u = \sum_{s \in \mathcal{S}} u_s(t) e^{is \cdot x}, \qquad f = \sum_{s \in \mathcal{S}} f_s(t) e^{is \cdot x}, \tag{2}$$

which are valid for all elements $u, f \in \mathbb{H}$. The solutions of equations (1) are naturally to satisfy the conditions

$$\mu u_s|_0 - u_s|_b = 0, \tag{Γ_s}$$

which follow from condition (Γ).

Lemma 1. *The $L - \Gamma$ problem is uniquely solvable for every element $f \in \mathbb{H}$ (or zero belongs to ρL) if and only if, under conditions (Γ_s), all the equations in the chain (1) have unique solutions, and there is a constant $c > 0$, independent of s, such that*

$$\|u_s\|_t \leq c \|f_s\|_t \quad \text{for every } s \in \mathcal{S} \tag{Φ_s}$$

(the norm in (Φ_s) refers to $\mathbb{H}_t$).

Proof. Sufficiency. Let all the equations (1) have unique solutions for all $f_s \in \mathbb{H}_t$ and let the inequalities (Φ_s) be satisfied uniformly in $s \in \mathcal{S}$. Then for sufficiently smooth solutions of the $L - \Gamma$ problem we have the inequality

$$\|u\| \leq c \|L u\|. \tag{Φ}$$

In fact, in this case the coefficients u_s in the representation (2) can be found automatically as solutions of (1) under the conditions (Γ_s), and it remains only to observe that

$$\|u\|^2 = (2\pi)^n \sum_{s \in \mathscr{S}} \|u_s\|_t^2.$$

Since the operator $\mathbf{L}$ is defined as the closure in $\mathbb{H}$ of the operation $\mathbf{L}(D)$, defined on smooth functions subject to the corresponding conditions, it follows from what has been said that (Φ) remains valid for every element $u \in \mathfrak{D}(\mathbf{L})$. The existence of the bounded operator $\mathbf{L}^{-1}$ follows from (Φ). Moreover, $\mathfrak{D}(\mathbf{L}^{-1}) = \mathbb{H}$, i.e. the $\mathbf{L} - \Gamma$ problem has a unique solution for all $f \in \mathbb{H}$. In fact, $\mathfrak{D}(\mathbf{L}^{-1})$ automatically contains all finite sums of the form (2), i.e. $\mathbf{L}^{-1}$ is defined on a dense set and therefore, since it is bounded, on the whole space $\mathbb{H}$.

Necessity. The failure of the unique solvability of any of the equations (1) implies the existence of a nontrivial solution $u_s(t)$ of the homogeneous equation; but then the corresponding function $u_s(t) e^{is \cdot x}$ would be a non-trivial solution of the homogeneous $\mathbf{L} - \Gamma$ problem.

If, however, (1) has a unique solution under conditions (Γ_s), but there is a sequence $\{f_{s_k}\}$ such that

$$\|u_{s_k}\|_t \geq k \|f_{s_k}\|_t, \qquad k = 1, 2, \ldots,$$

then $\mathbf{L}^{-1}$ exists, and is defined on a dense set (on finite sums of the form (2)), but is unbounded (it is enough to consider a sequence of right-hand sides for $(\mathbf{L})$ of the form $\{f_{s_k}(t) e^{is_k \cdot x}\}$). Since $\mathbf{L}^{-1}$ is closed, $\mathfrak{D}(\mathbf{L}^{-1})$ cannot be the whole space $\mathbb{H}$ in this case (Banach's theorem; 1.3 of Chapter I).

We now turn directly to the proof of the theorem. It will be presented in several steps.

First we observe that since $D_t - \mathbf{A} - \lambda = D_t - (\mathbf{A} + \lambda)$, where $\mathbf{A} + \lambda$ is again a Π-operator, it is sufficient to carry out the proof of the correspondence of spectra, as stated in the theorem, for the case $\lambda = 0$. We will denote the operator $\mathbf{T}(0)$ simply by $\mathbf{T}$.

Lemma 2. *If $\mu \in \rho \mathbf{T}$, then zero belongs to $\rho \mathbf{L}$.*

Proof. It is enough to establish that the hypotheses of Lemma 1 are satisfied for $\mu \in \rho \mathbf{T}$. By Proposition 3, §2, Chapter IV, the hypothesis of the lemma implies the existence of $\delta > 0$ such that

$$|\mu - \exp b \mathbf{A}(s)| \geq \delta \qquad \text{for all } s \in \mathscr{S}. \tag{3}$$

The unique solvability of all the equations (1) follows immediately from (3). It remains only to verify that the inequalities (Φ_s) are satisfied uniformly.

The representation that we have used for the solutions of (1), Formula (11), §2, Chapter III, does not explicitly show the dependence of $\mathbf{A}$ on s. Let $\mathbf{A} = r + iq$, where r and q are real. We have

$$|u|^2 \leq 2|\mu - e^{b\mathbf{A}}|^2 \left\{ |\mu|^2 \left| \int_0^t e^{(t-\tau)\mathbf{A}} f \, d\tau \right|^2 + e^{br} \left| \int_t^b e^{(t-\tau)\mathbf{A}} f \, d\tau \right|^2 \right\}.$$

For $r(s) = r = 0$ (as for every given value of r) it is evident that the inequalities (Φ_s) are satisfied with some constant c_0. Taking $r \neq 0$, we obtain

$$\left| \int_0^t e^{(t-\tau)A} f \, d\tau \right|^2 \leq \frac{e^{2tr} - 1}{2r} \|f\|_t^2 \equiv F_1(r) \|f\|_t^2,$$

$$\left| \int_t^b e^{(t-\tau)A} f \, d\tau \right|^2 \leq \frac{1 - e^{2(t-b)r}}{2r} \|f\|_t^2 \equiv F_2(r) \|f\|_t^2.$$

Define a number M by the equations $M = b^{-1}(1 + \ln|\mu|)$ for $\mu \neq 0$ and $M = 0$ for $\mu = 0$. Then for values of r satisfying the inequality $-\infty < r \leq M$, the numbers $F_1(r)$ and $e^{2br} F_2(r)$ are bounded by constants and it follows from (3) that the inequalities (Φ_s) are satisfied with a constant independent of s. If, however, $M < r < +\infty$, then if we note that in this case

$$|\mu - e^{bA}|^2 \geq (|\mu| - e^{br})^2 \geq \tfrac{1}{4} e^{2br},$$

we see that the numbers $e^{-2br} F_1(r)$ and $F_2(r)$ are again uniformly bounded, i.e. the inequalities (Φ_s) are again satisfied with a constant c_1 independent of s. $\square$

Lemma 3. *If $\mu \in P\sigma T$ then zero belongs to $P\sigma L$.*

Proof. Under the hypotheses of the lemma the equation $\mu = e^{bA(s_0)}$ is satisfied for some $s_0 \in \mathscr{S}$, and the function $\exp[i s_0 \cdot x + t A(s_0)]$ is a nontrivial solution of the homogeneous $L - \Gamma$ problem. $\square$

Lemma 4. *If $\mu \in C\sigma T$ then zero belongs to $C\sigma L$.*

Proof. As in the proof of Proposition 3, §2, Chapter IV, we see at once from the condition $\mu \notin P\sigma T$ that the operator L^{-1} exists and the domain $\mathfrak{D}(L^{-1})$ is dense in $\mathbb{H}$ (it necessarily contains all finite sums of the form (2)). Let us show that L^{-1} is unbounded.

By the hypothesis of the lemma there is a sequence $\{s^k\} \in \mathscr{S}$ such that

$$|\mu - \exp b A_k| = \varepsilon_k \to 0 \qquad \text{as } k \to \infty. \tag{4}$$

Here $A_k = A(s^k)$. Take the sequence of right-hand sides

$$f_k = \exp(i s^k \cdot x), \qquad \|f_k\|^2 = (2\pi)^n b. \tag{5}$$

Let us show that the norm of a solution u_k of the $L - \Gamma$ problem with a right-hand side of the form (5) increases unboundedly as $k \to \infty$. The form of u_k for the right-hand sides (5) is given by

$$u_k = u_k(t) \exp(i s^k \cdot x), \qquad \|u_k\|^2 = (2\pi)^n \|u_k(t)\|_t^2,$$

where $u_k(t)$ is found from formula (11), §2, Chapter IV, with $\lambda = A_k$, $f = 1$, i.e.

$$u_k(t) = (\mu - e^{bA_k})^{-1}\left(\mu\int_0^t e^{(t-\tau)A_k}\,d\tau + e^{bA_k}\int_t^b e^{(t-\tau)A_k}\,d\tau\right).$$

If we carry out the integrations and exclude the case $A_k = 0$ (which is always possible by passing, if necessary, to a subsequence), we obtain

$$u_k(t) = \frac{1}{A_k}\left(\frac{(\mu-1)\,e^{tA_k}}{\mu - e^{bA_k}} - 1\right). \tag{6}$$

If $\mu = 1$, then u_k is independent of t, $|A_k| \to 0$ by (4), and $|u_k| \to \infty$ as $k \to \infty$, so that the unboundedness of $\mathbf{L}^{-1}$ is established in this case.

If $\mu \neq 1$, then $|A_k| \geq \eta > 0$ for sufficiently large k, the term A_k^{-1} and the factor $(\mu - 1)$ can be dropped in (6), and it is enough to show that the norms of the functions

$$v_k = e^{tA_k}(\mu - e^{bA_k})^{-1}A_k^{-1}$$

increase unboundedly. Let $A_k = r_k + i q_k$, where r_k and q_k are real. Then

$$\|v_k\|_t^2 = \frac{e^{2br_k} - 1}{2r_k|A_k|^2\,|\mu - e^{bA_k}|^2}. \tag{7}$$

If $\mu = 0$, the right-hand side of (7) reduces to

$$\frac{1 - e^{-2br_k}}{2r_k(r_k^2 + q_k^2)}. \tag{8}$$

We now need to observe that the rate of growth (as $r \to -\infty$) of the numerator in (8) cannot be offset by the rate of growth of the denominator (because it contains the term $2r_k q_k^2$), since r and q are given polynomials. But then $\|v_k\|_t^2 \to \infty$ as $k \to \infty$.

The case $\mu = \infty$ can evidently be discussed similarly. If now $\mu \neq 0, 1, \infty$, then for sufficiently large k

$$0 < \eta_1 \leq |A_k^2| \leq \eta_2 < \infty.$$

Then, if we discard the factor $|A_k|^2$ from the denominator of (7) and observe that $|r_k|^{-1}|e^{2br_k} - 1| \geq \alpha > 0$ (since for sufficiently large k we have $|e^{2br_k} - 1| \geq \delta_1 > 0$, $r_k > -N$), we see that $\|v_k\|_t^2$ increases together with $|\mu - e^{bA_k}|^{-2}$. $\square$

The theorem stated above follows immediately from Lemmas 1-3. Numerous applications of the theorem will be discussed in §3 during our study of various kinds of operators L. Meanwhile we merely make some remarks to clarify the fundamental position of the Cauchy problem for equation (L). As above, we use the notation $\mu = \infty$ to indicate the boundary condition $u|_{t=0} = 0$.

Proposition 1. *For $\mu = 0$ or ∞, and every Π-operator* **A**, *the point spectrum of* **L** *is empty.*

The conclusion follows from the fact that the points 0 and ∞ cannot belong to $P\sigma$**T**. □

Proposition 1 means that the Cauchy problem for (L) always has a unique solution.

Proposition 2. *If $\mu = 0$ or ∞, either all finite points of the complex plane* $\mathbb{C}$ *belong to ρ**L** (if 0 or ∞ belongs to ρ**T**), or else all points of $\mathbb{C}$ belong to* $C\sigma$**L** *(if 0 or ∞ belongs to $C\sigma$**T**).*

The conclusion follows from the fact that the point zero (or infinity) belongs to ρ**T** or $C\sigma$**T** independently of the replacement of **A**(s) by **A**(s) $+ \lambda$. □

Therefore when $\mu = 0$ or ∞ the operator **L** is a $q\,C$-operator (3.5, Chapter I) whenever $\mathbf{L}^{-1}$ is bounded. There is also an evident connection between Proposition 2 and the existence of the energy inequalities for solutions of the Cauchy problem (1.1, Chapter III), whose validity does not depend on the form of the minor part.

In accordance with the remarks after the statement of the general theorem, we note that when $\mu \neq 0, \infty$ the set $P\sigma$**L** can be described by the formula

$$-\mathbf{A}(s) + b^{-1}[\ln|\mu| + i\arg\mu + 2k\pi i], \quad s\in\mathscr{S},\ k = 0, \pm 1, \pm 2, \dots.$$

Here $\sigma\mathbf{L} = \overline{P\sigma\mathbf{L}}$, $C\sigma\mathbf{L} = \sigma\mathbf{L}\backslash P\sigma\mathbf{L}$.

§2. The Operator $D_t - A$; Special Boundary Conditions

Before using the results of §1 for analyzing the properties of some specific partial differential operations (which we do in §3), we consider some questions of a general nature. We preserve the notation and hypotheses of §1.

It follows immediately from the theorems of §1 that whenever the resolvent set of $\mathbf{T} = \exp(b\mathbf{A})$ is not empty (**A** is a Π-operator; $\mathbf{T}: \mathbb{H}_x \to \mathbb{H}_x$), there are proper operators generated by the operation

$$\mathbf{L}(D) \equiv D_t - \mathbf{A}(-iD) \tag{1}$$

and boundary conditions on t of the form

$$\mu u|_{t=0} - u|_{t=b} = 0. \tag{2}$$

However, by analogy with Proposition 4, §2, Chapter IV, it is easy to establish the following fact.

Proposition 1. *For $n=1$ the set ρT is always nonempty; for $n>1$ there exist operators A such that σT fills the entire complex plane $\mathbb{C}$.*

Proof. For $n=1$ the conclusion is trivial. For the second part of the conclusion it is enough to observe that the spectrum of T automatically fills the complex plane $\mathbb{C}$ if the spectrum of A has this property. At the same time, the set of values of the function $\exp(b A(s))$, $s \in \mathscr{S}$, and of the function (of s and k)

$$\exp[b A(s) + i 2\pi k], \quad s \in \mathscr{S}, \ k = 0, \pm 1, \pm 2, \ldots, \tag{3}$$

are obviously the same. But for

$$b/2\pi \, A(s) = s_1 + \alpha s_2 + i \beta s_2^2$$

(α, β irrational) the set of values of the function

$$b \, A(s) + 2k \pi i = 2\pi \left[\frac{b}{2\pi} A(s) + i k \right]$$

(s and k as in (3)) is dense in $\mathbb{C}$, as follows from the proof of Proposition 4, §2, Chapter IV. $\square$

Thus when $n \geq 2$ there is a Π-operator A such that for every choice of μ in (2) each point of the complex plane belongs either to the point spectrum or to the continuous spectrum of the corresponding operator $L: \mathbb{H} \to \mathbb{H}$. Nevertheless, if we turn to the argument used in §3, Chapter II, it is easy to see that for every Π-operator A there must exist boundary conditions for t (i.e., conditions on $u(x,t)$ for $t=0$ and b) such that the operator $L: \mathbb{H} \to \mathbb{H}$ determined by them is proper. Let us consider this situation.

Proposition 2. *For every Π-operator A, the operator $L_0: \mathbb{H} \to \mathbb{H}$ defined by an operation of the form (1) with the following conditions on t:*

$$u|_{t=0} = u|_{t=b} = 0 \tag{4}$$

has a bounded inverse L_0^{-1}.

Proof. It is enough to observe that under conditions (4), for $u \in \mathfrak{D}(L_0)$, satisfying the equation $L_0 u = f$, we may define $u_s(t)$ in the representation (2), §1, by either of the equations

$$u_s = \int_0^t e^{(t-\tau)A_s} f_s \, d\tau, \qquad u_s = - \int_t^b e^{(t-\tau)A_s} f_s \, d\tau. \tag{5}$$

It follows from the first equation that

$$|u_s(t)|^2 \leq \|f_s\|_t^2 \int_0^t e^{2(t-\tau)\mathrm{Re}\, A_s}\, d\tau \tag{6}$$

and we can obtain a similar inequality by starting from the first equation in (5). Therefore if we use the first equation (5) for $\mathrm{Re}\, A_s \leq 0$ and the second for $\mathrm{Re}\, A_s > 0$, we see at once that the inequalities (Φ_s), § 1, are satisfied uniformly for $s \in \mathscr{S}$, which implies the existence of the bounded operator L_0^{-1}. □

It is now easy to guess that the "t-minimal" operator L_0 defined in Proposition 2 must play the role of the minimal operator in § 3, Chapter II. As the "t-maximal" operator we select the operator $\tilde{L}$ generated by the operation $L(D)$ on functions that are free of any conditions on t. Since, in the situation under consideration, the strong and weak definitions of the operators L_0 and $\tilde{L}$ are trivially equivalent, then if we use the constructions of § 3, Chapter II, we immediately have the following results.

Proposition 3. *We have the equation* $\Re(\tilde{L}) = \mathbb{H}$. □

Proposition 4. *There is a proper operator* L *such that* $L_0 \subset L \subset \tilde{L}$. □

Now, however, in contrast to the general situation in Chapter II, for every Π-operator we can describe explicitly the class of boundary conditions on t that determine L. Before we proceed to this description, we notice two propositions that follow from the results of § 1, Chapter III, and from § 1 of the present chapter.

Let $L_s \colon \mathbb{H}_t \to \mathbb{H}_t$, $t \in (0, b)$, be an operator generated by the ordinary differential operation $D_t - A(s)$. The set of these operators for $s \in \mathscr{S}$ evidently generates an operator $L \colon \mathbb{H} \to \mathbb{H}$ (with a corresponding domain), if we define $Lu = \sum_{s \in \mathscr{S}} L_s u_s(t) e^{is \cdot x}$, where we used the representation (2), § 1, for u.

Proposition 5. *Under the same hypotheses, every proper restriction of* $\tilde{L}$ *is determined by a set* $\{L_s\}$, $s \in \mathscr{S}$, *of proper restrictions of the maximal operators* $\tilde{L}_s$ *generated on* $(0, b)$ *by the ordinary differential operations* $D_t - A(s)$.

The set of proper restrictions $\{L_s\}$, $s \in \mathscr{S}$, *determines the proper restriction* L *of* $\tilde{L}$ *if and only if the norms of* $L_s^{-1} \colon \mathbb{H}_t \to \mathbb{H}_t$ *are uniformly bounded for* $s \in \mathscr{S}$.

Proposition 6. *Proposition 5 remains valid if "proper restriction" is replaced by "proper operator."* □

We now find a method for giving boundary conditions in t that generate a proper operator $L \colon \mathbb{H} \to \mathbb{H}$ for an arbitrary Π-operator A. For a given A, we divide $\mathscr{S}$ into two subsets $\mathscr{S}^+$, $\mathscr{S}^-$, by setting

$$s \in \mathscr{S}^- \quad \text{if} \quad \mathrm{Re}\, A(s) \leq 0; \qquad s \in \mathscr{S}^+ \quad \text{if} \quad \mathrm{Re}\, A(s) > 0.$$

This partition induces a partition of $\mathbb{H}_x$ into a sum of orthogonal subspaces: $\mathbb{H}_x = \mathbb{H}_x^+ + \mathbb{H}_x^-$, where $\mathbb{H}_x^+$ (or $\mathbb{H}_x^-$) is the closed linear span of the vectors $\exp(i s \cdot x)$, $s \in \mathscr{S}^+$ (or $s \in \mathscr{S}^-$). Let μ^- and μ^+ be the projections on $\mathbb{H}_x^-$ and $\mathbb{H}_x^+$.

Theorem. *Specifying the domain of the operator* $\mathbf{L}$ *by the conditions*

$$\mu^- u|_{t=0} - \mu^+ u|_{t=b} = 0 \tag{7}$$

determines a proper operator $\mathbf{L}: \mathbb{H} \to \mathbb{H}$ *for every Π-operator* $\mathbf{A}$.

Proof. It is sufficient to observe that conditions (7) are equivalent to requiring the definition of $u_s(t)$ (in the solution of the equation $\mathbf{L}u = f$) by the first formula (5) for $s \in \mathscr{S}^-$ and by the second for $s \in \mathscr{S}^+$. Because of inequalities of the form (6), this guarantees that the inequalities (Φ_s) of §1 are satisfied uniformly for $s \in \mathscr{S}$, whence the conclusion of the theorem follows.

Remark. Conditions (7) provide the simplest example of the use of a pseudodifferential operator (of order zero) for the description of a boundary value problem. One of the first examples of conditions of this type was suggested in [27] for the discussion of a boundary value problem in a half space for an ultrahyperbolic operator. As will be seen in Chapter VI (in the discussion of operator equations of second order) in the case of an ultrahyperbolic operator one can also get along with ordinary conditions of the type of (2). At the same time, as follows from the preceding construction, in some cases the use of special conditions like (7) is unavoidable.

Proposition 7. *Under our hypotheses, every point λ of the (finite) complex plane $\mathbb{C}$ belongs to $\rho \mathbf{L}$ when the operator $\mathbf{L}$ is defined by conditions (7). If in addition $|\operatorname{Re} \mathbf{A}(s)| \to \infty$ as $|s| \to \infty$, $\mathbf{L}^{-1}$ is a Volterra operator.*

In fact, passing from the operator $\mathbf{A}$ to $\mathbf{A} + \lambda$, with any given finite λ, leads only to replacing the inequality $\operatorname{Re} \mathbf{A}(s) \leq 0$ (for $s \in \mathscr{S}^-$) by $\operatorname{Re} \mathbf{A}(s) \leq M$ or to a similar shift of the inequality $\operatorname{Re} \mathbf{A}(s) > 0$. This evidently does not affect the validity of inequalities of the form (6).

The conditions in the second part of the conclusion lead to a decrease in $\|u_s\|_t$ which guarantees the complete continuity of $\mathbf{L}^{-1}$ (cf. the proof of Proposition 5, §2, Chapter IV). $\square$

Therefore a problem defined by conditions (7) has, in some sense, properties close to those of a Cauchy problem (considered in detail in the last part of §1). The corresponding operator $\mathbf{L}$ is (in the terminology of 3.5, Chapter I) a $q C$-operator.

§3. The Operator $D_t - \mathbf{A}$; Classification

We continue to use the notation and definitions of §1. Various properties of the operators $\mathbf{L}$ that we consider arise, naturally, from the differences in the properties of the Π-operators $\mathbf{A}$ used in their definitions. In 2.2, Chapter IV, we listed some basic types of operators $\mathbf{A}$, and the present section is directly connected to that subsection. However, in the passage to the operation $\mathbf{L}(D)$ there arises in the first instance a difference in the roles of the real part $\mathbf{R}(s)$ and the imaginary part $\mathbf{Q}(s)$ of the polynomial $\mathbf{A}(s) = \mathbf{R} + i \mathbf{Q}$.

We recall that we say that the real polynomial $\mathbf{R}(s)$ has the C-*property* if the limit

$$\lim_{|s| \to \infty, \, s \in \mathscr{S}} |\mathbf{R}(s)| = \infty \tag{C}$$

exists, and has the B-*property* if one of the inequalities

$$\inf_{s \in \mathscr{S}} \mathbf{R}(s) \geq -M > -\infty, \quad \sup_{s \in \mathscr{S}} \mathbf{R}(s) \leq M < +\infty \tag{B}$$

is satisfied.

The polynomial $\mathbf{R}(s)$ has the *strong* C-property (or B-property) if (C) (or one of the inequalities (B)) is satisfied for every $s \in \mathbb{R}^n$. These definitions were discussed in 2.2, Chapter IV.

An equation (L):

$$\mathbf{L} u \equiv (D_t - \mathbf{A}) u = f, \tag{L}$$

considered together with conditions (Γ):

$$\mu u|_{t=0} - u|_{t=b} = 0, \tag{Γ}$$

is called an $L - \Gamma$ problem, and is said to be *regular* if the corresponding operator $\mathbf{L}$ is proper (i.e., $0 \in \rho \mathbf{L}$ or $\mu \in \rho \mathbf{T}$).

To the polynomials $\mathbf{R}(s)$ and $\mathbf{Q}(s)$ there correspond selfadjoint Π-operators $\mathbf{R}$ and $\mathbf{Q}$, and we may speak of the spectra $\sigma \mathbf{R}$, $\sigma \mathbf{Q}$, etc. If $\mathbf{R}(s)$ has the B-property, then $\sigma \mathbf{R}$ must lie on a ray of the real axis (or $\sigma \mathbf{A}$ must lie in the corresponding half plane); this is a necessary and sufficient condition for the regularity of either the direct or the inverse Cauchy problem for (L). Here the absence of the strong B-property for $\mathbb{R}$ indicates that the Cauchy problem is ill-posed in the noncompact case (i.e., if $\mathbb{T}^n$ is replaced by $\mathbb{R}^n$), corresponding to a continuous spectrum (the polynomial $s_1^2(s_2 + \alpha)^2 + s_1$ with nonintegral α has the B-property but not the strong B-property).

The additional presence of the C-property is a necessary condition for the stability of the regular problem with respect to perturbations of $\mathbf{A}$. If $\mathbf{R}$ has the C-property but not the B-property, there is always a sufficient supply of regular values $\mu \in \rho \mathbf{T}$, $\mu \neq 0, \infty$. Here the absence of the strong C-property indicates that the corresponding problem is specific for the compact case. Finally, the absence of both B and C properties for $\mathbf{R}$ can lead to situations in which $\rho \mathbf{L}$ is empty.

Passing to the consideration of the skew-symmetric part $\mathbf{Q}$ of the operator $\mathbf{A}$, we say that a regular $L - \Gamma$ problem is *stable* with respect to perturbations of the operator $\mathbf{A}_0 = \mathbf{R}_0 + i\mathbf{Q}_0$ if there is a $\delta > 0$ such that for every operator of the form

$$D_t - (\mathbf{A} + \varepsilon\,\mathbf{R}_0 + i\eta\,\mathbf{Q}_0)$$

the $L - \Gamma$ problem remains regular for $|\varepsilon| + |\eta| \leq \delta$ (in contrast to 2.2, Chapter IV, it is now more convenient to distinguish between perturbations of $\mathbf{R}$ and of $\mathbf{Q}$).

In explaining the influence of the properties of $\mathbf{Q}$ on the spectrum of $\mathbf{T}$, a fundamental role is played by the circles O_s:

$$O_{s_0} = \{z:\ |z| = \exp \mathbf{B}\,\mathbf{R}(s_0),\ s_0 \in \mathscr{S}\}.$$

The spectrum $P\sigma\mathbf{T}$ evidently belongs to the union $O_{\mathscr{S}}$ of the circles O_s. Among the O_s it is natural to single out the *circles $\hat{O}_s$ of instability*, which have the property that the sum of the multiplicities of the points $P\sigma\mathbf{T}$ that lie on the $\hat{O}_s$ is infinite.

For $\mu \notin O_{\mathscr{S}}$ the $L - \Gamma$ problem is trivially regular and, in addition, stable with respect to arbitrary perturbations of $\mathbf{Q}$. For $\mu \in \rho\,\mathbf{T}$, $\mu \in \hat{O}_S$, where O_s is not a circle of instability, the problem is stable with respect to sufficiently small perturbations of $\mathbf{Q}$. If, however, a regular problem corresponds to a value $\mu \in \hat{O}_S$, then naturally an arbitrarily small perturbation of $\mathbf{Q}$ can produce an operator for which $\mu \in P\sigma\mathbf{T}$ (typical situation: the Dirichlet problem for the wave equation; see below).

When $\mathbf{R}$ has the C-property, the question of the stability of a problem for which $\mu \notin O_s$ reduces to the question of the stability of the spectrum of $\mathbf{R}$, which we discussed in 2.2, Chapter IV. Furthermore, as in that subsection, it is most essential to consider the relation between the groups $\mathscr{S}'$ and $\mathscr{S}''$ on which $\mathbf{R}(s)$ and $\mathbf{Q}(s)$ depend.

It should be noted that the distinction between the partial dependence of $\mathbf{Q}$ and its independence (2.2, Chapter IV) will not be essential in what follows, and we shall call $\mathbf{Q}$ *independent* in either case. In describing the properties of $\mathbf{L}$ it is sufficient to distinguish three kinds of polynomials $\mathbf{Q}$:

1. $\mathbf{Q} \equiv 0$;
2. $\mathbf{Q} \not\equiv 0$, $\mathbf{Q}$ depends completely on $\mathbf{R}$;
3. $\mathbf{Q} \not\equiv 0$, $\mathbf{Q}$ is independent.

It is useful to notice that when $\mathbf{R}$ has property (C) a circle of instability exists only for operators with independent skew-symmetric part $\mathbf{Q}$.

Taking account of the preceding discussion, we can classify operators $\mathbf{A}$ (and consequently $\mathbf{L}$) by assigning to each $\mathbf{A}$ a pair of symbols $(\mathbf{X}, \mathbf{Y})$ and supposing that the first symbol describes the properties of $\mathbf{R}$; the second, the properties of $\mathbf{Q}$. A complete table constructed on this principle would contain more than 20 types of operators $\mathbf{A}$. It is, however, more expedient to restrict ourselves to the analysis of a number of the most typical examples.

It is convenient to focus our attention on the strong C- and B-properties. Therefore, in the pair (X, Y), we shall replace X by C (or B) if R has the strong C-property (or strong B-property), by $\sim$ if R does not satisfy either of the conditions B, C, and by zero if $R \equiv 0$. The symbols 0, D, I in place of Y mean that the corresponding operators are identically zero, depend on R, or are independent.

Type $(C, 0)$; $n > 1$. The simplest example is the operator generated by the operation

$$L(D) \equiv D_t \pm (D_1^2 + D_2^2 - 1),$$

which, from the classical point of view, is parabolic. The operator A is selfadjoint, with spectrum on a ray of the real axis and having a single limit point, ∞ or $-\infty$. The operator A^{-1} is completely continuous. Among the $L - \Gamma$ problems either the direct or the inverse Cauchy problem is always regular.

Type $(C, 0)$; $n = 1$. This case is in a certain sense singular, since when $n = 1$ there is no implication C $\Rightarrow$ B. The classical example illustrating the situation that arises yields

$$L(D) \equiv D_t + i D_x$$

(Cauchy-Riemann operator). Both the direct and inverse Cauchy problems are irregular. Regular problems are stable.

Type (C, I). The simplest example is

$$L(D) \equiv D_t \pm D_x^2 + D_x^{2p+1}.$$

In the general case, if the spectrum of R lies on the positive semiaxis, the spectrum of A is in the right-hand half plane, and in each bounded domain in $\mathbb{C}$ the number of points of the spectrum is bounded.

Type (C, I). The simplest example is

$$L(D) \equiv D_t \pm D_1^2 + D_2.$$

The spectrum of A again lies in a half plane, but can be dense on certain vertical lines. All circles O_s are circles of instability. If the set of points of $P\sigma T$ on $\hat{O}_s$ is finite, then each one has infinite multiplicity; if the multiplicities are finite, then $P\sigma T$ is dense in $\hat{O}_s$. One case can turn into the other under arbitrarily perturbations of Q or of the parameter b.

Type $(B, 0)$. Example:

$$L(D) \equiv D_t + (D_1 - D_2)^2.$$

The operator A is selfadjoint; the spectrum of A lies on a semi-axis. The principal difference from type $(C, 0)$ consists in the presence of limit points of the spectrum on bounded intervals of the real axis and in the behavior under perturbations of R.

Type (B, D). Example:

$$\mathbf{L}(D) \equiv D_t + (D_1 - D_2)^2 + D_1^{2p+1}.$$

The difference from type (C, D) is the occurrence of limit points in finite domains of $\mathbb{C}$ (there are circles of instability).

Type (B, I). Example:

$$\mathbf{L}(D) \equiv D_t + (D_1 - D_2)^2 + D_3.$$

All circles O_s are circles of instability; the principal difference from type (C, I) is the behavior under perturbations of $\mathbf{R}$.

Type $(\sim, 0)$. Example:

$$\mathbf{L}(D) \equiv D_t + D_1^2 - D_2^2.$$

The operator $\mathbf{A}$ is selfadjoint, but its spectrum is not semi-bounded. An arbitrarily small perturbation of $\mathbf{R}$ (or of the parameter b) can induce a transition from a pure point spectrum for $\mathbf{A}$ (containing points of infinite multiplicity) to a spectrum filling the whole real axis.

Type $(\sim, \mathrm{D})$. The operator $\mathbf{A}$ is obtained in this case by adding to a "bad" $\mathbf{R}$ a dependent skew-symmetric operator. The spectrum of $\mathbf{A}$ is dispersed over the whole complex plane, but does not necessarily fill the whole plane.

Type $(\sim, \mathrm{I})$. This type includes the example in §2 of an operator $\mathbf{A}$ such that the spectrum of $\mathbf{L}$ fills the whole plane $\mathbb{C}$, even when $n = 2$.

Type (0, I). The simplest example is

$$\mathbf{L}(D) \equiv D_t + D_x.$$

The spectrum of $\mathbf{A}$ lies on the imaginary axis; this example is the simplest hyperbolic operator. The only circle (the unit circle) O_s is a circle of instability. Both the direct and inverse Cauchy problems are regular. The situation is unstable with respect to perturbations of $\mathbf{A} \equiv i\mathbf{Q}$ by a symmetric operator $\mathbf{R}$.

It is interesting to notice that the same type contains the Schrödinger operator, generated by the operation

$$\mathbf{L}(D) \equiv D_t \pm i D_x^2$$

(D_x^2 can be replaced by the Laplacian Δ). The relationship of the Schrödinger operator to the simplest hyperbolic operator is one of the classical (but not trivial) facts of theoretical physics.

We shall confine ourselves to these examples. In conclusion we notice the following fact.

Proposition. *An operation* $L(D)$ *of the type we have considered is hyperbolic (see* [9]) *if and only if the operator* A *has type* $(C, 0)$.

§4. Operators not Solvable for D_t

We now turn to discussing questions of what specific features differentiate an operator equation

$$L u \equiv A_0 D_t u + A_1 u = f \tag{1}$$

from the equations considered in §§ 1-3, i.e. from equations in which $A_0 = 1$. The phenomena that arise correspond, in a certain sense, to phenomena that are inherent in the classical theory of problems with boundary value on characteristics.

In contrast to the considerations of the preceding sections, we turn to the consideration (within the scope of our scheme) of one of the classical partial differential equations connected with the operation

$$L(D) \equiv D_x D_t.$$

Applying our scheme means that we want to consider the equation

$$L_\lambda u \equiv D_x D_t u - \lambda u = f \tag{2}$$

in the rectangle $V = (0 < x < 2\pi) \times (0 < t < b)$, with periodicity in x and the conditions

$$\mu u|_{t=0} - u|_{t=b} = 0 \tag{Γ}$$

on t. Supposing that $f \in \mathbb{H}(V)$, we define a solution of the problem $(2) - \Gamma$ to be an element $u \in \mathbb{H}$ that satisfies (2), with the operator $L_\lambda : \mathbb{H} \to \mathbb{H}$ thought of as the closure in $\mathbb{H}$ of the operation $L_\lambda(D)$, defined initially on functions possessing a continuous derivative $D_x D_t$ and satisfying the above boundary conditions in the classical sense.

If now $u(t, x)$ is a sufficiently smooth function that satisfies (2) and the boundary conditions in the classical sense, we can use our standard representation

$$u(t, x) = \sum u_s(t) e^{i x s}, \quad s \in \mathscr{S},$$

and the analogous representation for $f(t, x)$, to obtain the chain of equations

$$i s D_t u_s - \lambda u_s = f_s \tag{3}$$

for u_s, where each u_s satisfies the supplementary conditions (Γ). Let us consider the question of constructing a solution of the original problem by starting from the chain of equations (3).

First let $\mu \neq 0, \infty$. Then for $\lambda \neq 0, s \neq 0$, the function $u_s(t)$ is determined by the condition

$$\mu - \exp(b\lambda/(is)) \neq 0 \tag{4}$$

by formula (11), §2, Chapter III. Condition (4) automatically implies that inequalities (Φ_s) of §1 are satisfied uniformly for $s \in \mathscr{S}$ (corollary of the one-dimensionality of $\mathbf{A}_0$). However, if $\lambda \neq 0$ has the property that condition (4) is violated for some $s \in \mathscr{S}$, the corresponding function

$$\exp\left\{\frac{\lambda}{is} t + i s x\right\}$$

is an eigenfunction.

For $\lambda \neq 0, s = 0$, the value of u_0 is determined by the equation

$$u_0(t) = -\lambda^{-1} f_0(t), \tag{5}$$

but in this case $u_0(t)$ will not, generally speaking, satisfy conditions (Γ) (unless $f_0(t)$ is specifically required to satisfy this condition).

Finally, if $\lambda = 0$ and $s = 0$ the function $u_0(t)$ is arbitrary, and the corresponding equation in (3) is solvable only under the additional condition $f_0(t) = 0$.

For $\mu = 0, \lambda \neq 0, s \neq 0$, all the equations (3) trivially have unique solutions, and the case $\lambda \neq 0, s = 0$ again leads to (5). The existence of the difference between the condition $\mu = 0$ and the "regular" conditions (Γ) is now explained by the fact that the point $\lambda = 0$ belongs to $P\sigma\mathbf{L}$, and the corresponding space of eigenfunctions is infinite-dimensional and consists of all functions of the form $u(t)$ (more precisely, of the form $u(t) \otimes 1$; see §1, Chapter IV).

The case $\mu = \infty$ can be discussed similarly.

The preceding discussion lets us give a complete description of the operator $\mathbf{L} \colon \mathbb{H} \to \mathbb{H}$ generated by the operation $L(D)$ and conditions (Γ).

Theorem. *For $\mu \neq 0, \infty$ the spectrum of the operator $\mathbf{L} \colon \mathbb{H} \to \mathbb{H}$ defined above is a pure point spectrum, and the set of eigenvalues is described by the equation*

$$\lambda_{s,k} = \frac{is}{b}[\ln|\mu| + i \arg\mu + 2k\pi i]; \quad s, k = 0, \pm 1, \dots \tag{6}$$

When $\mu = 0$ or ∞ the only point of the spectrum of $\mathbf{L}$ is $0 \in P\sigma\mathbf{L}$; the corresponding space of eigenfunctions consists of all functions of the form $u(t) \otimes 1$.

Proof. After the preceding discussion it remains only to add that for a given regular value λ (not in the spectrum as described) the proof given in §1 of the existence and uniqueness of a generalized solution when inequalities (Φ_s), §1, are satisfied uniformly for $s \in \mathscr{S}$ is evidently applicable to the present situation. In fact, the definition of $u_0(t)$ by solutions (5) for the construction of a smooth approximating sequence of functions that satisfy (Γ) is unchanged, since in the sequence of approximating functions (in $\mathbb{H}(V)$) we may always suppose that the right-hand sides are functions $f_{0,i}(t)$ (or even $f_i(t, x)$) that satisfy conditions (Γ). $\square$

Remark. As in the corresponding remark on the fundamental theorem of §1, we observe that formula (6), corresponding to case $\mu \neq 0, \infty$, illustrates the classical proposition "the spectrum of a product of commuting operators is the direct product of their spectra."

This remark shows, on the other hand, that our approach (method of defining solutions of (2)) is natural from the point of view of including the problems that we consider in the theory of operators on $\mathbb{H}$.

It is clear that the properties that we have established for the Cauchy problem are connected with the spectral properties of the operators D_x and D_t for corresponding boundary conditions, but no abstract theorem applicable to this case is known.

Now, in order to apply (to some extent) our discussion of equation (2) to the study of the general entity (1), we adjoin a minor term to (2), i.e. we replace (2) by the equation

$$L_\lambda u \equiv (D_x D_t + a_1 D_t + a_2 D_x - \lambda) u = f, \tag{7}$$

where a_1 and a_2 are constants. Then (3) is replaced by

$$(is + a_1) D_t u_s + (a_2 i s - \lambda) u_s = f_s. \tag{8}$$

We can again apply the preceding scheme to equations (7) and (8). Here there springs into view the very strong influence of the "minor" terms on the nature of the solutions of the $(7)-(\Gamma)$ problem; in particular, on the nature of the solvability of the Cauchy problem, which is now connected with the presence of a point spectrum for the Cauchy conditions also (in problem $(2)-(\Gamma)$).

The phenomenon that we have mentioned, of the strong influence of the "minor part", is well known in the classical theory of the so-called "characteristic problems" (see the remark at the beginning of this section).

With corresponding complications, our scheme evidently carries over also to the general operator equation (1).

§ 5. Differential Properties of the Solutions of Operator Equations, and Related Questions

We return again to the consideration of the simplest equation

$$L u \equiv (D_t - A) u = f \tag{L}$$

under the conditions

$$\mu u|_{t=0} - u|_{t=b} = 0, \tag{Γ}$$

retaining the hypotheses of §1, and elucidate first of all the differential properties of the solutions of regular (§3) $L - \Gamma$ problems.

By the study of differential properties we understand arguments similar to those used for the proof of Proposition 6, §2, Chapter IV. The question of differential properties is closely related to the question of the complete continuity of $\mathbf{L}^{-1}$, and the possession by $\mathbf{L}^{-1}$ the property of complete continuity lets us include in the discussion of the operation $\mathbf{L}(D) + \mathbf{M}(D)$, obtained by taking account of the perturbation of $\mathbf{L}(D)$ by "minor terms" with coefficients depending on t and x.

Thus let $u(x, t)$ be a solution of the regular $L - \Gamma$ problem. Then it follows immediately from the inequalities used in the proof of Lemma 2, §1, that the coefficients $u_s(t)$ of our standard representation (2), §1, of $u(t, x)$ satisfy the inequality

$$\|u_s(t)\|_t \le \frac{c}{|r(s)|} \|f_s(t)\|_t, \quad s \in \mathscr{S}, \ r(s) \neq 0, \tag{1}$$

where the constant c is independent of s. For values of s for which $r(s) = 0$, equation (1) is simply replaced by

$$\|u_s(t)\|_t \le c \|f_s(t)\|_t,$$

where c is again independent of s.

Inequality (1) is evidently sharp, i.e. we cannot replace $r(s)$ by any polynomial $r_1(s)$ such that

$$|r|/|r_1| \to 0 \quad \text{as } |s| \to \infty.$$

The following proposition follows immediately from what we have said.

Proposition 1. *The solution of the regular $L - \Gamma$ problem both possesses a generalized derivative $D_t u \in \mathbb{H}$ and belongs, for every $t \in [0, b]$, to the domain of the operator $\mathbf{A}$: $\mathbb{H}_x \to \mathbb{H}_x$, if and only if there is a constant $M < \infty$ such that*

$$|\mathbf{A}(s)|/|r(s)| \le M \quad \text{uniformly for } s \in \mathscr{S}. \tag{2}$$

Proof. It is enough to use the equation $D_t u_s = f_s + \mathbf{A}(s) u_s$ and inequality (1), and to observe that $\mathbf{A}u \in \mathbb{H}$ if and only if $D_t u \in \mathbb{H}$. $\square$

Proposition 2. *When (2) is satisfied, the corresponding operator $\mathbf{L}^{-1}$: $\mathbb{H} \to \mathbb{H}$ (the $L - \Gamma$ problem is regular) is completely continuous if and only if*

$$\lim_{|s| \to \infty} |\mathbf{A}(s)| = \infty.$$

The proof of Proposition 2, using (1) and (2), is a repetition of the proof of Proposition 5, Chapter IV. □

Example. For the operator L: $\mathbb{H}\to\mathbb{H}$ generated under our hypotheses by the regular problem for the operation

$$\mathbf{L}(D)\equiv D_t+i(D_1+D_2),$$

the solution of equation (L) will have a derivative $D_t u\in\mathbb{H}$, but $\mathbf{L}^{-1}$ will not be a CC operator.

The possession by $\mathbf{L}^{-1}$ of the property of complete continuity lets us discuss perturbations of L that are qualitatively different from the perturbations, discussed above, of A by various Π-operators.

Let $\mathbf{M}(D)$ have the form (in the notation of Chapter II)

$$\mathbf{M}(D)u\equiv\sum_{|\alpha|\leq p}a_\alpha(t,x)D^\alpha u,\tag{3}$$

where a_α is continuous in t and has enough smoothness with respect to x_k so that the operation $\mathbf{M}^t(D)$ is defined. Let L: $\mathbb{H}\to\mathbb{H}$ be a given operator of the form (L), satisfying the hypotheses of Proposition 2. Let $W^{1,p+1}$ be the Hilbert space of functions that have, in the region $V_t\times V$ under consideration (§1), the generalized derivative D_t and all (generalized) derivatives with respect to x_k (§7, Chapter II) up to order $p+1$. If it follows from $u\in\mathfrak{D}(L)$ that $u\in W^{1,p+1}$ (which is defined because of the validity of (2) and the properties of $A(s)$ given in Proposition 6, §2, Chapter IV), then, in the first place, the operator M: $\mathbb{H}\to\mathbb{H}$ generated by the operation (3) is defined on $u\in\mathfrak{D}(L)$ in a natural way (since the generalized derivatives that appear in (3) belong to $\mathbb{H}$). In the second place, not only is $\mathbf{L}^{-1}$: $\mathbb{H}\to\mathbb{H}$ a CC operator, but so is $\mathbf{L}^{-1}\mathbf{M}$, as follows immediately from the proposition of §7, Chapter II.

We now consider the operator equation

$$(\mathbf{L}+\mathbf{M})u=f,\tag{4}$$

where L and M are taken in the sense described above. Here it is convenient to observe that because of our conventions the operator on the left side of (4) can immediately be defined in the usual way, i.e. as the closure in $\mathbb{H}$ of the operation $\mathbf{L}(D)+\mathbf{M}(D)$ considered initially on smooth functions of period 2π in x_k and subject to conditions (Γ) in t.

Equation (4) is equivalent, under our hypotheses, to

$$(1+\mathbf{L}^{-1}\mathbf{M})u=\mathbf{L}^{-1}f=g.\tag{5}$$

We may suppose, without loss of generality, that $1\in\rho(\mathbf{L}^{-1}\mathbf{M})$. In fact, in the contrary case we would replace M by $\kappa\mathbf{M}$, where κ is arbitrarily close to 1 ($\mathbf{L}^{-1}\mathbf{M}$ is a CC operator, and 1 can only be an isolated eigenvalue). If $1 \in \rho(\mathbf{L}^{-1}\mathbf{M})$ then (5), and consequently (4), is solvable for every $g, f \in \mathbb{H}$, i.e. the operator $(\mathbf{L}+\mathbf{M})^{-1}$ exists and is bounded.

Proposition 3. *Under our hypotheses,* $(\mathbf{L}+\mathbf{M})^{-1}$ *is completely continuous.*

In fact, $\mathbf{L}+\mathbf{M}=\mathbf{L}(1+\mathbf{L}^{-1}\mathbf{M})$ and $(\mathbf{L}+\mathbf{M})^{-1}=(1+\mathbf{L}^{-1}\mathbf{M})^{-1}\mathbf{L}^{-1}$, where $(1+\mathbf{L}^{-1}\mathbf{M})^{-1}$ is bounded, and $\mathbf{L}^{-1}$ is completely continuous. $\square$

If we now apply the theorems of 2.3 of Chapter I on the relation between the spectra of an operator and its inverse, and the nature of the spectral properties of CC operators (3.1 and 3.2 of Chapter I), we can make corresponding statements about the spectrum of $\mathbf{L}+\mathbf{M}$. Moreover, if we consider along with $\mathbf{L}+\mathbf{M}$ the adjoint operator $(\mathbf{L}+\mathbf{M})^*$, we can establish analogs of the Fredholm theorems for this pair. It is sufficient to pass to the operators $(\mathbf{L}+\mathbf{M})^{-1}$ and $(\mathbf{L}+\mathbf{M})^{-1*}$ by using the equation $(\mathbf{L}+\mathbf{M})^{-1*}=(\mathbf{L}+\mathbf{M})^{*-1}$ (1.3 of Chapter I) and the connection between the spectra of an operator and its inverse.

The greatest interest, however, is presented by the possibility of replacing $(\mathbf{L}+\mathbf{M})^*$ by $\mathbf{L}^t+\mathbf{M}^t$, where the latter operator is defined by starting from the operation $\mathbf{L}^t(D)+\mathbf{M}^t(D)$ in the same way in which $\mathbf{L}+\mathbf{M}$ was defined, with (Γ) replaced by the condition

$$v|_{t=0} - \bar{\mu} v|_{t=b} = 0. \qquad (\Gamma_t)$$

Proposition 4. *Under our hypotheses the Fredholm theorems are valid for the pair of operators* $\mathbf{L}+\mathbf{M}$, $\mathbf{L}^t+\mathbf{M}^t$.

The proof evidently reduces to the verification of the equation

$$(\mathbf{L}+\mathbf{M})^* = \mathbf{L}^t+\mathbf{M}^t. \qquad (6)$$

However, the equation $\mathbf{L}^*=\mathbf{L}^t$ follows immediately from results that we have obtained previously (constancy of coefficients, periodicity conditions). Consequently the regularity of the value μ automatically implies the regularity of conditions (Γ_t). In addition, the operator $(\mathbf{L}^t)^{-1}\mathbf{M}^t$: $\mathbb{H} \to \mathbb{H}$ is completely continuous, and the equation $(\mathbf{L}^t+\mathbf{M}^t) v=h$ under the supplementary hypothesis $1 \notin \sigma[(\mathbf{L}^t)^{-1}\mathbf{M}^t])$ is uniquely solvable for every $h \in \mathbb{H}$. This ensures (6.4 of Chapter II) the validity of equation (6). $\square$

In the framework of these constructions we may also consider the case $1 \in \sigma(\mathbf{L}^{-1}\mathbf{M})$. In this case, because of the complete continuity of $\mathbf{L}^{-1}\mathbf{M}$, the point 1 must belong to the point spectrum. Then the kernel of the operator must be finite-dimensional and the condition for the solvability of (5) is written in the usual way, using the orthogonality of the right-hand side to the elements of the kernel of the adjoint operator.

What we have said above is usually summed up in the statement, "Equation (4) is normally-solvable." The supplementary hypothesis $1 \notin \sigma(\mathbf{L}^{-1}\mathbf{M})$ is, as a rule, not discussed.

§ 6. Some Operators with Variable Coefficients in the Principal Part

In this section we discuss some examples of operators of the simplest type

$$\mathbf{L} \equiv D_t - \mathbf{A},$$

for which functions of x or t appear in the definition of $\mathbf{A}$ (in the "principal part").

As a first example we consider the operator generated by the operation

$$\mathbf{L}(D) u \equiv D_t u - \mathbf{A}(x) u, \tag{1}$$

where the function $\mathbf{A}(x)$ is continuous for $x \in [0, 2\pi]$.

As before, we suppose that $t \in [0, b]$. We shall be interested in the operator $\mathbf{L} \colon \mathbb{H} \to \mathbb{H}$, $\mathbb{H} = \mathbb{H}_t \otimes \mathbb{H}_x$, defined as the closure in $\mathbb{H}$ of the operation (1), considered initially on smooth functions subject to the condition

$$\mu u|_{t=0} - u|_{t=b} = 0. \tag{Γ}$$

As before, in studying the spectrum of $\mathbf{L}$ it is enough to consider only the case $\lambda = 0$ (the case of general λ reduces to this by replacing $\mathbf{A}$ by $\mathbf{A} + \lambda$). The operation of multiplying by a function generates an operator in $\mathbb{H}$ with continuous spectrum, and it is natural to expect that the spectrum of $\mathbf{L}$ is obtained by "smearing out" the point spectrum of D_t (induced by conditions (Γ), $\mu \neq 0, \infty$), since in (1) we are dealing with a difference of commuting operators.

In fact, we have the following proposition.

Proposition 1. *Under the condition*

$$\mu - \exp[b \, \mathbf{A}(x)] \neq 0, \quad x \in [0, 2\pi], \tag{2}$$

the equation

$$\mathbf{L} u = f \tag{3}$$

is uniquely solvable for every $f \in \mathbb{H}$. If condition (2) is violated at a finite number of points $x_j \in [0, 2\pi]$, $j = 1, \ldots, N$, then zero belongs to $C \sigma \mathbf{L}$.

Proof. The first part of the proposition follows because under condition (2) the solution of (3) is given by formula (11), § 2, Chapter III, with λ replaced by $\mathbf{A}(x)$.

For the proof of the second part, it is sufficient to observe that the same formula also yields the solution of (3) when condition (2) is violated at the points x_j, if the right-hand side f satisfies the supplementary condition

$$f(t, x) \equiv 0 \quad \text{for } |x - x_j| \leq \varepsilon, \; j = 1, \ldots, N, \; t \in [0, b]. \tag{4}$$

The set of $f \in \mathbb{H}$ that satisfy (4) (for all $\varepsilon > 0$) is dense in $\mathbb{H}$. The operator L^{-1}, defined on this dense set, is evidently unbounded (its norm increases unboundedly as ε in (4) tends to zero). $\square$

The pathological case in which (2) is violated at an infinite number of points will not be of interest here. When $\mu = 0, \infty$ (direct and inverse Cauchy problems) equation (3) is always uniquely solvable.

The second of the examples of interest will show that even for an operation of the form

$$\mathbf{L}(D) \equiv D_t - \varphi(t)\,\mathbf{A},$$

where $\mathbf{A}$ is a Π-operator and $\varphi(t)$ is a smooth function, the proposition on the possibility of always being able to give a corresponding proper operator $\mathbf{L}$ with the aid of a condition on t (for $t = 0, b$) becomes invalid (in contrast to Propositions 2-4, § 2).

Let us consider the simplest case $\mathbf{A} \equiv D_x^2$, $\varphi(t) = 2t - b$, i.e. the operation

$$\mathbf{L}(D) \equiv D_t - (2t - b)\,D_x^2, \quad t \in [0, b]. \tag{5}$$

Remark. The operation (5) corresponds, evidently, to inverse heat conduction for $t = 0$ and to direct heat conduction for $t = b$; this indicates the character of the example.

Let us recall that the t-minimal operator generated by an operation $\mathbf{L}(D)$ of the type under consideration is the operator determined by the conditions

$$u|_{t=0} = u|_{t=b} = 0.$$

In order to show that there does not exist a proper operator generated by (5), it is sufficient to verify the following proposition (cf. Proposition 2, § 2).

Proposition 2. *For the t-minimal operator $\mathbf{L}_0$ generated by (5), the inverse $\mathbf{L}_0^{-1}$: $\mathbb{H} \to \mathbb{H}$ is unbounded.*

Proof. If we use our usual representation $u = \sum u_s(t)\,e^{isx}$ and the similar representation for $f(t, x)$, the solution of (3) with the operator $\mathbf{L}$ determined by (5) will satisfy

$$D_t u_s + (2t - b)\,s^2\,u_s = f_s. \tag{6}$$

We take a sequence of functions $\{f_{(k)}\}$ of the form

$$f_{(k)}(t, x) = k^2 (2t - b)\,e^{ikx}, \quad k = 1, 2, \ldots.$$

The corresponding solutions of (3) will, as follows from (3), be functions $u_{(k)}$ of the form

$$u_{(k)} = [e^{k^2 t(b-t)} - 1] e^{ikx},$$

that evidently satisfy the conditions $u_{(k)}|_{t=0} = u_{(k)}|_{t=b} = 0$. Then $\|u_{(k)}\| / \|f_{(k)}\| \to \infty$ as $k \to \infty$. □

Finally we notice that the preceding scheme lets us study a model of the simplest transition problems. Let, for example, $V_t = (-b < t < b)$,

$$\mathbf{L} \equiv D_t - \mathbf{A}_\sigma,$$

where $\sigma = 1$ if $t \in (-b, 0)$ and $\sigma = 2$ if $t \in (0, b)$, and A_1, A_2 are Π-operators. If we adjoin to conditions of the form (Γ) (for $t = \pm b$) the condition that $u(t, x)$ is continuous at $t = 0$, then the properties of the operator $\mathbf{L} \colon \mathbb{H} \to \mathbb{H}$, $\mathbb{H} = \mathbb{H}_t(V_t) \otimes \mathbb{H}_x$, defined in the corresponding way, will determine the spectra of the operators A_σ and the choice of the parameter μ in (Γ).

Considerations of this kind evidently admit many variants.

§7. Concluding Remarks

As we remarked in the introduction, this chapter is, in a certain sense, central to our whole study, and it is natural to supplement it with a number of remarks on the structure of our exposition as a whole.

Our discussion of operator equations of the first order shows that our approach allows us to study models of various situations that arise in the investigation of boundary value problems for linear partial differential operations in a bounded domain. Sections 1–3 and 5 were devoted to the general case, and 4–6 to problems of a special nature, the number of which could have been greatly increased (by considering, for example, equations with a small parameter, with degenerate coefficients, with moving boundaries, etc.). However, we have preferred to concentrate on a basic method of investigation.

In the next chapter this method will be applied to operator equations of second order (where we restrict ourselves to the "main case") and some remarks will be made on its application to equations of arbitrary order.

Chapter VII is devoted to the study (in the same circle of ideas) of the general question of the existence of the proper operator generated by an arbitrary partial differential operation with constant coefficients in a bounded region.

From the general point of view (2.2, Chapter I) our construction reduces to the study of the properties of the operator $\mathbf{L}^{-1}$ (determined by the $L - \Gamma$ problem) defined as a function of the operator $\mathbf{A}$ (or of commuting operators $\mathbf{A}_0, \mathbf{A}_1$), depending on an auxiliary parameter $t \in (0, b)$. The study is facilitated by supposing that $\mathbf{A}$ is an M-operator.

Without the preceding assumptions, the construction of a solvable operator L^{-1} (in the analogous situation) is unavoidably connected with the use
of the resolvent of A, for which we must use special supplementary requirements that replace the corresponding requirements on the spectrum. As one
may readily guess, constructions that use the resolvent of A are considerably
less simple than the case considered in the present chapter.

Constructions that are useful in effecting the transition within the framework of our methods will be presented in Chapter VIII. There we shall be
able to see that for the operation

$$L(D) \equiv D_t + a D_x, \qquad t \in [0, b], \ x \in [0, 2\pi],$$

where $a > 0$ is a real number, the Cauchy condition on t:

$$u|_{t=0} = 0$$

determines a proper operator, not only under conditions of periodicity (or
regularity) in x, but also under the condition $u|_{x=0} = 0$.

Chapter VI
Operator Equations of Higher Order

§0. Introductory Remarks

With the hypotheses and notation of the introductory section of Chapter V, we can write the general differential-operator equation of mth order in the form

$$\mathbf{L}u \equiv (\mathbf{A}_0 D_t^m + \mathbf{A}_1 D_t^{m-1} + \ldots + \mathbf{A}_m) u = f. \tag{1}$$

Although, in principle, the method of Chapter V (under the hypothesis that $\mathbf{A}_k$ is a $\varPi$-operator or an M-operator) is also directly applicable to (1), a number of technical points make it necessary to change the plan of the investigation in an essential way. It is enough to remark that already for $m > 2$ there is no satisfactory (and for $m > 4$, no) explicit representation for the roots of the characteristic equation

$$\sum_{j=0}^{m} \mathbf{A}_{m-j} z^j = 0, \quad \mathbf{A}_k = \text{const}, \tag{2}$$

in terms of the coefficients. This affects the method of using the solutions of the auxiliary ordinary equation of (1) (in which $\mathbf{A}_k$ are numbers) which arises in our discussion. Moreover, the boundary conditions of general form, defined by proper operators generated by the ordinary differential operations (1) contain at least m^2 essential parameters (§§ 2,3, Chapter III), and it has not been possible to obtain a sufficiently transparent characterization of the properties of the spectra of the corresponding operators for all possible choices of the boundary conditions.

It is therefore natural that in discussing equations of the form (1) we have to restrict ourselves to special types of operators and special classes of boundary conditions (at $t = 0, b$). In this chapter we discuss the case $m = 2$, in view of its comparative simplicity and its direct connection with the classical entities of the theory of boundary value problems for the equations of mathematical physics, and some special classes of boundary value problems for particular cases of equation (1).

We also note that the consideration of the characteristic equation of the form (2), connected with (1), in which the A_k are operators, leads in a natural way to the concept of a pencil of operators – one of the popular entities of classical spectral theory (see [40]).

§ 1. Second-Order Operator Equations

1.0. Preparatory Remarks. In the study of an operator equation of the form (1), §0, with $m=2$, our main attention, as in Chapter V, is focused on the case $A_0 \equiv 1$. The passage to the general case can be carried out as in §4, Chapter V.

Retaining the hypotheses of §0, Chapter V, we write our equation in the form

$$\mathbf{L}u \equiv (D_t^2 + 2\mathbf{B}D_t - \mathbf{A})u = f. \tag{1}$$

If $\mathbf{A}$ and $\mathbf{B}$ are Π-operators (or M-operators) the basic machinery for studying (1) is the formula that gives the solution of the corresponding ordinary differential equation in which $\mathbf{A}$ and $\mathbf{B}$ are constants. The essential difference from Chapter V is that (as was mentioned above) the general boundary conditions determining a proper operator generated by the ordinary differential operation (1) contains at least four essential parameters (§3, Chapter III) and a sufficiently transparent description of the properties of the spectrum of the operator, under all types of boundary conditions, is not available.

We also note that since §1 of this chapter corresponds to the whole of Chapter V, the subsections here correspond approximately to the sections of the preceding chapter.

1.1. Elementary Formulas. If

$$k^2 + 2\mathbf{B}k - \mathbf{A} = 0 \tag{2}$$

is the characteristic equation of (1), and k_1, k_2 are its roots,

$$k_{1,2} = -\mathbf{B} \pm \sqrt{\mathbf{B}^2 + \mathbf{A}}, \tag{3}$$

then when $k_1 \neq k_2$ the solution of the Cauchy problem

$$u|_{t=0} = u'_t|_{t=0} = 0 \tag{4}$$

has the form

$$u(t) = \int_0^t \frac{e^{k_1(t-\tau)} - e^{k_2(t-\tau)}}{k_1 - k_2} f(\tau)\, d\tau \equiv I(t)f, \tag{5}$$

and the general solution of (1) can be represented by the formula

$$u(t)=c_1\,e^{k_1 t}+c_2\,e^{k_2 t}+I(t)f. \tag{6}$$

For multiple roots $k_1=k_2=k$ the representation (5) is to be replaced by (5'):

$$u(t)=\int_0^t e^{k(t-\tau)}(t-\tau)f(\tau)\,d\tau\equiv\hat{I}(t)f, \tag{5'}$$

and (6) by (6'):

$$u(t)=c_1\,e^{kt}+c_2\,t\,e^{kt}+\hat{I}(t)f. \tag{6'}$$

It is useful to notice that $\lim\limits_{k_1\to k_2=k} I(t)=\hat{I}(t).$

1.2. A General Method. The basic discussion in this chapter will concern (as in Chapter V) the case when **A** and **B** in (1) are Π-operators, and we retain all the previously introduced notation. In studying the properties of equation (1) we again use the representation of u and f by series of the form (2), §1, Chapter V, and consider the resulting chain of ordinary differential equations. The solutions of these equations will be given by formulas (6) and (6') with values of the roots $k_1(s)$, $k_2(s)$ that depend on $A(s)$ and $B(s)$, $s\in\mathscr{S}$, i.e. on the values of polynomials generated by the Π-operators **A** and **B**.

The proper operators generated under our hypotheses by the operation $L(D)$ are determined by various supplementary conditions that make it possible to determine the constants $c_{1,s}$ and $c_{2,s}$ in (6) and (6'). The availability of the inequalities

$$\|u_s(t)\|_t\le c\,\|f_s(t)\|_t \tag{Φ_s}$$

uniformly for $s\in\mathscr{S}$ will (as in Chapter V) automatically yield an existence and uniqueness theorem for the generalized solution of (1), where the operator on the left is again understood as the closure of the differential operation defined initially on smooth functions belonging to various special linear manifolds.

In the following discussion we shall, as a rule, confine the analysis to the "general case" $k_1\neq k_2$, i.e. to formulas (5) and (6) that yield families of solutions $u_s(t)$. The supplementary study of the case of multiple roots presents no difficulty.

1.3. The Cauchy Problem. Supposing that **A** and **B** in (1) are Π-operators, we first consider the classical Cauchy problem.

Theorem 1. *If there is a sequence $\{s_{(i)}\}\subset\mathscr{S}$ such that we can select a corresponding sequence of roots $k(s_{(i)})=k_i$ of (2) that satisfy the requirement $\operatorname{Re}k_i\to+\infty$ as $i\to\infty$, then for the operator $L\colon \mathbb{H}\to\mathbb{H}$ generated by the* *operation* (1) *and the Cauchy conditions* (4), *the continuous spectrum fills the complex plane* $(C\sigma L = \mathbb{C})$.

If on the contrary there is a constant M *such that* $\mathrm{Re}\, k_{1,2}(s) \leq M < +\infty$ *for every* $s \in \mathcal{S}$, *then* $\rho L = \mathbb{C}$, *i.e. every point of the complex plane belongs to the resolvent set of the operator.*

Proof. For given $\mathbf{A}(s)$ and $\mathbf{B}(s)$ let the hypotheses of the first part of the theorem be satisfied and let $\{s_i\} \subset \mathcal{S}$ be the corresponding sequence. Taking the sequence of right-hand sides $f_{(i)} = e^{s_{(i)} \cdot x}$ (for which $\|f_{(i)}\|$ is evidently constant), we immediately find that the solution $u_{(i)}$ of equation (1) obtained from (5) or (5') for $f(\tau) \equiv 1$ satisfies

$$\|u_{(i)}\| \to \infty \quad \text{as} \quad i \to \infty$$

(this is obvious for (5'); for (5) we must also observe that, for example, when $\mathrm{Re}\, k_1 > M > 0$ we have $\|u\| \geq M' > 0$ uniformly for $k_2 \neq k_1$, where $M' \to +\infty$ as $M \to +\infty$).

When given functions $\mathbf{A}(s)$ and $\mathbf{B}(s)$ satisfy the hypotheses of second part of the theorem, the solution is again given by a chain of ordinary differential equations and (Φ_s) is satisfied uniformly for $s \in \mathcal{S}$. This, as we noticed above, immediately yields the existence and uniqueness theorem for the solution of the operator equation for an arbitrary right-hand side in $\mathbb{H}$.

It remains only to remark that if the hypotheses of the theorem are satisfied for a polynomial $\mathbf{A}(s)$ they are evidently satisfied when $\mathbf{A}(s)$ is replaced by $\mathbf{A}(s) + \lambda$, where λ is any complex number. $\square$

A theorem analogous to Theorem 1 evidently holds for the inverse Cauchy problem:

$$u|_{t=b} = u'_t|_{t=b} = 0. \tag{7}$$

We present a brief statement.

Theorem 1'. *If there is a sequence* $\{s_{(i)}\} \subset \mathcal{S}$ *such that the roots* $k_i = k(s_{(i)})$ *of equation* (2) *satisfy the condition* $\mathrm{Re}\, k_i \to \infty$ *as* $i \to \infty$ *then the operator* $(L) -$ (7) *has* $C\sigma L = \mathbb{C}$. *If* $\mathrm{Re}\, k_{1,2}(s) \geq -M > -\infty$ *then* $\rho L = \mathbb{C}$.

Theorems 1 and 1' evidently contain the standard result on the ill-posed nature of the Cauchy problem for the elliptic equation of second order ($\mathbf{B} = 0$, $\mathbf{A}(s) = \sum_k s_k^2$; $C\sigma L = \mathbb{C}$) and its correctness for the hyperbolic case ($\mathbf{B} = 0$, $\mathbf{A} = -\sum_k s_k^2$; $\rho L = \mathbb{C}$). Moreover, since we may always suppose $\mathbf{B} = 0$ for a second-order equation, it follows immediately from (3) that the direct (4) or inverse (7) Cauchy problems are always simultaneously either correct or ill-posed. The example of the operation

$$\mathbf{L}(D) \equiv D_t^2 - 2D_1^2 D_t + D_1^4 - D_2^2; \quad \mathbf{B} = s_1^2, \quad \mathbf{A} = -s_1^4 - s_2^2, \quad \mathrm{Re}\, k_1 \leq 0, \quad \mathrm{Re}\, k_2 \leq 0,$$

shows that in general this is not always so.

According to Theorem 1 and 1', an operator that is defined by Cauchy conditions and is proper is necessarily a $q\,C$-operator (3.5, Chapter I). Here, however, in contrast to the classical case of the differential operations of mathematical physics, it is not necessarily a C-operator (for example, when $\mathbf{A} = -(s_1 + s_2)^2$ and $\mathbf{B} = 0$).

1.4. Existence of Proper Operators. As the next step in the study of equation (1) we prove the following proposition which is analogous to Proposition 4, §2, Chapter V: we will show that even in the case when the spectra of the Π-operators $\mathbf{A}$ and $\mathbf{B}$ fill the complex plane, there are proper operators generated by (1) and certain boundary conditions at $t=0$ and b. In the course of the proof the principal difference of the case $m=2$ from the case $m=1$ will stand out clearly. The plan of the proof is altered correspondingly.

Lemma. *There is a constant $M>0$ such that for every pair $\mathbf{A}$, $\mathbf{B}$ of complex numbers the proper operator $\mathbf{L}:\mathbb{H}_t \to \mathbb{H}_t$ generated by the operation (1) can be defined (by corresponding boundary conditions at $t=0$ and b) so that*

$$\|\mathbf{L}^{-1}\| \le M. \tag{8}$$

Proof. As boundary conditions determining the required proper operator we take conditions of the form

$$\mu u|_0 - u|_b = 0, \qquad \mu u_t'|_0 - u_t'|_b = 0 \tag{9}$$

(in other words, we consider D_t^2 as the square of the operator D_t determined by the first condition (9)) and solve the equation

$$\mathbf{L}u \equiv (D_t^2 + 2\mathbf{B}D_t - \mathbf{A})u(t) = f(t), \tag{10}$$

by expanding u and f in eigenfunctions of the operator D_t, which in this case have the form

$$e^{\lambda_p t}, \qquad \lambda_p = b^{-1}\{\ln|\mu| + i\arg\mu + 2p\pi i\}, \qquad p = 0, \pm 1, \dots .$$

If $u = \sum u_p e^{\lambda_p t}$ then for the solution u of equation (10) the coefficients u_p are determined by the equations

$$u_p(\lambda_p^2 + 2\mathbf{B}\lambda_p - \mathbf{A}) = f_p, \qquad p = 0, \pm 1, \pm 2, \dots .$$

To prove the lemma it is now sufficient to verify that for some $M_0 > 0$ and all $\mathbf{A}$ and $\mathbf{B}$ the number μ in (9) can be chosen so that

$$|\lambda_p^2 + 2\mathbf{B}\lambda_p - \mathbf{A}| \ge M_0, \tag{11}$$

simultaneously for all p. But since

$$|z^2 + 2\mathbf{B}z - \mathbf{A}| = |z - k_1||z - k_2|,$$

where k_1, k_2 are given complex numbers, and all the eigenvalues λ_p lie on the line $\operatorname{Re} z = b^{-1} \ln|\mu|$, it is evidently possible to choose μ to have the required properties (even for an arbitrarily preassigned M_0). $\square$

Remark. We are dealing in essence with the study of an operator function $z^2 + 2\mathbf{B}z - \mathbf{A}$ of a complex parameter z. As we mentioned in §0, such a function is an example of a pencil of operators.

We noticed above that the principal difference from the discussion in Chapter V is that it is impossible to confine ourselves to the values $\mu = 0$ or $\mu = \infty$ (Cauchy conditions) in (9). This impossibility follows from (5) and the corresponding formula for the inverse Cauchy problem: as $k_1 \to \infty$ and $k_2 \to -\infty$ the norm of $\mathbf{L}^{-1}$ determined by either formula increases unboundedly.

Now let $\mathbf{A}$ and $\mathbf{B}$ be any Π-operators and $\mathbf{A}(s)$, $\mathbf{B}(s)$ the corresponding polynomials. Again using the representation

$$u(t, x) = \sum u_s(t) e^{is \cdot x} \tag{12}$$

and choosing a number M_0, we select the parameter $\mu = \mu_s$ in the boundary conditions (9) for $u_s(t)$ (determined in accordance with equation (10) with $\mathbf{A} = \mathbf{A}_s$, $\mathbf{B} = \mathbf{B}_s$) so that (11) is satisfied. Supposing now that a set $\{\mu_s\}$, $s \in \mathscr{S}$, is given, with all μ_s satisfying the preceding condition, we set (for a given $f \in \mathbb{H}$) $\mathbf{L}^{-1} f = u$, where u is given by (12), in which (with obvious notation) $u_s = \mathbf{L}_s^{-1} f_s$. The operator $\mathbf{L}^{-1}$ defined in this way is a bounded operator with domain $\mathbb{H}$.

Let us formulate this result as a theorem.

Theorem 2. *For any operation $\mathbf{L}(D)$ of the form* (1), *the operator* $\mathbf{L}$: $\mathbb{H} \to \mathbb{H}$ *defined by the equations*

$$\mathfrak{D}(\mathbf{L}) = \mathbf{L}^{-1}\mathbb{H}, \qquad \mathbf{L}u = \mathbf{L}(\mathbf{L}^{-1}f) = f,$$

where $\mathbf{L}^{-1}$, is the operator described above, is a proper operator generated by the operation (1). $\square$

It should be noted that here, again in contrast to the case $m = 1$, the operator $\mathbf{L}$ is not in general a qC-operator (not excluding the possibility of a point spectrum). It is, however, not difficult to give examples of operations for which our construction does provide a proper operator which is a qC-operator (this is the case if in (9) we always take $\mu = 0$ or $\mu = \infty$), and for which neither the direct nor the inverse Cauchy problem determines such an operator. It is enough to take, for example,

$$\mathbf{L}(D) \equiv D_t^2 + (D_1^2 - D_2^2) D_t.$$

1.5. The Dirichlet Problem. If the boundary conditions for t that determine a proper operator can be chosen independently of s and guarantee the validity of (11) for all $\mathbf{A}(s)$ and $\mathbf{B}(s)$ then we obtain a proper operator $\mathbf{L}$: $\mathbb{H} \to \mathbb{H}$ determined by conditions that we shall call *standard* (in contrast to the conditions which were called *special* in §2, Chapter V, and in which the boundary conditions on t for determining $u_s(t)$ depend on s).

Conditions (9) do not include, however, such classical standard conditions as, for example, the Dirichlet conditions

$$u|_{t=0} = u|_{t=b} = 0. \tag{13}$$

We discuss these separately.

Theorem 3. *The point spectrum of the operator* $\mathbf{L}$: $\mathbb{H} \to \mathbb{H}$ *generated by the operation* (1) *and conditions* (13) *consists of the points of the complex plane* C *of the form*

$$-p^2 \frac{\pi^2}{b^2} - \mathbf{B}^2(s) - \mathbf{A}(s), \qquad p = \pm 1, \pm 2, \dots, s \in \mathscr{S}.$$

The complement $\mathbb{C} \backslash P\sigma\mathbf{L}$ *of the point spectrum belongs entirely either to the resolvent set* $\rho\mathbf{L}$ *or to the continuous spectrum* $C\sigma\mathbf{L}$ *of the operator* $\mathbf{L}$.

Proof. Having the equation

$$\lambda + p^2 \frac{\pi^2}{b^2} + \mathbf{B}^2(s) + \mathbf{A}(s) = 0 \tag{14}$$

satisfied for some integer $p \neq 0$ (and given λ) is equivalent to having $e^{k_1(s)b} = e^{k_2(s)b}$ satisfied for some $s \in \mathscr{S}$, where k_1 and k_2 are found from (3) with $\mathbf{B} = \mathbf{B}(s)$ and $\mathbf{A} = \mathbf{A}(s) + \lambda$. But then the function

$$u_v(t, x) = \exp(is \cdot x + k_1(s)t) - \exp(is \cdot x + k_2(s)t)$$

is an eigenfunction of our problem, associated with the eigenvalue λ.

Having (14) satisfied with $p = 0$ corresponds to the case of multiple roots: $k_1 = k_2 = k$. But in this case, as follows from (5') and (6'), the operator $\mathbf{L}_s^{-1}$ always exists (in (6')

$$c_1 = 0, \quad c_2 = -b^{-1} e^{-kb} \int_0^b e^{k(b-\tau)}(b-\tau) f \, d\tau, \text{ i.e. } \lambda \notin P\sigma\mathbf{L}).$$

If $\lambda \notin P\sigma\mathbf{L}$, the solutions $u_s(t)$ of the chain (1.2) of ordinary differential equations with conditions (13) (assuming $k_1 \neq k_2$) are given by the formula

$$u_s(t) = \mathbf{L}_s^{-1} f_s \equiv I(t) f_s - \frac{e^{k_1 t} - e^{k_2 t}}{e^{k_1 b} - e^{k_2 b}} I(b) f_s. \tag{15}$$

If $|\operatorname{Re} k_1(s)|$, $|\operatorname{Re} k_2(s)|$ are uniformly b ounded for $s \in \mathscr{S}$, or if the condition

$$\lim(\operatorname{Re} k_1(s)/\operatorname{Re} k_2(s)) = -1 \qquad (16)$$

is satisfied as $|\operatorname{Re} k_1(s)| \to \infty$ and $|\operatorname{Re} k_2(s)| \to \infty$, then we can obtain from (15) (or the corresponding formula in the case of multiple roots) the chain of inequalities (Φ_s) of 1.2, which ensure the existence and uniqueness of the solution of the equation (1) for every right-hand side $f \in \mathbb{H}$.

If these conditions are not satisfied, then, if we choose a sequence $\{s_{(j)}\} \subset \mathscr{S}$ for which (16) fails, and the corresponding sequence $\{u_{(j)}\}$ of solutions of (1) for the right-hand sides $f_{(j)} = \exp(i s_{(j)} \cdot x)$ (which corresponds to taking $f_{s(j)} = 1$ in (15)), we obtain

$$\|u_{(j)}\| / \|f_{(j)}\| \to \infty \qquad \text{as } j \to \infty,$$

i.e. the unboundedness of $\mathbf{L}^{-1}$.

It remains only to remark that whether or not the preceding conditions are satisfied is evidently independent of the change of $\mathbf{A}(s)$ to $\mathbf{A}(s) + \lambda$. $\quad\square$

Let us notice some corollaries of the theorem proved above. Condition (16) is trivially satisfied for equations of the second order with real coefficients. For the Laplace operator ($\mathbf{B} = 0$, $\mathbf{A} = \sum s_k^2$) the whole spectrum lies on the negative half-axis; for $\lambda \notin P\sigma \mathbf{L}$ the operator $\mathbf{L}^{-1}: \mathbb{H} \to \mathbb{H}$ is, as is easily shown, completely continuous, and its norm is determined by the distance from λ to the nearest point of the spectrum.

For the classical hyperbolic operation ($\mathbf{B} = 0$, $\mathbf{A} = -\sum s_k^2$) the Dirichlet conditions also determine a proper operator (proper in the wide sense; 4.1, Chapter III), i.e. the complement of the point spectrum is the resolvent set. However, here there arises the typical instability phenomenon for the point spectrum: when $\lambda = 0$, $A = s^2$ ($n = 1$), whether or not (14) is satisfied for certain integers p ($p \neq 0$) and s: $p\pi = bs$, depends on whether or not b and π are commensurable, and zero can become a point of the point spectrum (or of the resolvent set) by an arbitrarily small change in b (or replacement of $\mathbf{A}$ by $(1 + \varepsilon)\mathbf{A}$ with an arbitrarily small ε).

Finally, for the operation

$$\mathbf{L}(D) \equiv D_t^2 - 2 D_x^2 D_t + D_x^4 - D_y^2,$$

considered at the end of 1.3, the complement of the point spectrum belongs to the continuous spectrum, and the resolvent set is empty, i.e. condition (13) does not, in this case, determine a proper operator.

We emphasize that the possibility of having such a situation arise (a situation in which the complement of the point spectrum is not dense in $\mathbb{C}$) is a special feature of what are called *splitting* conditions (i.e., conditions which do not contain a connection between the values of the unknown function at $t=0$ and $t=b$ (see [18]), to which the Dirichlet conditions belong. This possibility is excluded for conditions (9) with $\mu \neq 0$, ∞.

1.6. Using the Standard Conditions. It follows from the preceding discussion that neither the Cauchy problem nor the Dirichlet problem generates proper operators for, for example, operations such as

$$\mathbf{L}(D) \equiv D_t^2 + aD_x^{2n+1}, \qquad n=0,1,\ldots, \tag{17}$$

$$\mathbf{L}(D) \equiv D_t^2 + D_1^2 - D_2^2 - D_3^2. \tag{18}$$

At the same time, the spectra of the corresponding Π-operators $\mathbf{A}$ ($\mathbf{B}=0$) are rather sparse, and there is no reason to suppose that in order to obtain proper operators generated by (17) or (18) we have necessarily to resort to the general construction of 1.4.

In fact, even such standard classical conditions as periodicity ($\mu_s = \mu = 1$ in (9)) do generate operators for (17) and (18) that are proper in the wide sense (the inconvenience of the condition of periodicity in all the variables is that zero is always a point of the spectrum of $\mathbf{L}$ if $\mathbf{A}(0)=0$).

Moreover, there is an infinite set of possible choices of real numbers μ in (9) for which

$$|\ln|\mu| + 2\pi i p \pm b\sqrt{A(s)}| \geq M_0 \geq 0$$

for $p=0, \pm 1, \pm 2, \ldots$, and for every $s \in \mathscr{S}$, if

$$\mathbf{A}(s) = a(is_1)^{2r+1}, \qquad \mathbf{A}(s) = s_1^2 - s_2^2 - s_3^2,$$

as in examples (17) and (18).

In addition, as was shown in [24] and as follows at once from the explicit formulas for finding u_s from the equations $\mathbf{L}u_s = f_s$, instead of using conditions (9) for describing proper operators $\mathbf{L}$ generated by the operations (17) and (18) we may use conditions that are closer to the classical ones:

$$u|_{t=0} = 0, \qquad \mu u'_t|_{t=0} - u|_{t=b} = 0. \tag{19}$$

We should note that conditions (19) are not regular in Birkhoff's sense (§3, Chapter III) for the operation $D_t^2 - \mathbf{A}$, and the discussion in 1.3 (using an expansion in eigenfunctions of the operator $\mathbf{L}: \mathbb{H} \to \mathbb{H}$) is not applicable in this case. Instead, we must investigate explicit formulas for $\mathbf{L}_s^{-1}$, similar to those in 1.3 and 1.5 (or in Chapter V).

A more detailed analysis (see [40]) shows that the necessity of introducing two boundary conditions of nonlocal character (like the second condition (19)) inevitably arises in the case when the real parts $\operatorname{Re} k_1(s)$, $\operatorname{Re} k_2(s)$ of both roots $k_1(s)$, $k_2(s)$ can simultaneously assume arbitrarily large positive or arbitrarily large negative values (which is evidently possible only when $\mathbf{B} \neq 0$).

1.7. A Concluding Remark. With this we conclude the discussion of operator equations of second order. It remains to note that a (detailed) analysis of the complete continuity of the operator **L**, or the differential properties of the solutions, or a discussion of operators not solvable for D_t^2, or of different approaches to the classification of equations (1) according to the properties of the polynomials $A(s)$ and $B(s)$, can be carried out in the framework of the constructions used in Chapter V. We shall not pursue these topics here.

§2. Operator Equations of Higher Order $(m > 2)$

2.0. Preparatory Remarks. As we noted in the introduction to this chapter, the transition to the study of the general operator equation of the form (1), §0, is accompanied by the emergence of new difficulties. In this section we present some examples of equations of higher order for which there is an approach that lets us obtain rather clear results, and make some comments on the general case. Our discussion will be of the nature of a survey. To give detailed proofs would require us to overcome a whole series of technical difficulties connected with the investigation of formulas that yield solutions of boundary value problems for chains of ordinary differential equations

$$L_s(D_t) u_s = f_s, \quad s \in \mathscr{S},$$

that arise in using an approach that follows the general method of 1.2, §1. For proofs we refer to [38] and [39].

We note only that these proofs depend on a detailed study of the properties of the Green functions which let one write down representations for the solutions of general boundary value problems for ordinary differential equations (formula (11), §2, Chapter III is also a representation of a solution by means of a Green function).

2.1. Binomial Equations. A binomial equation is of the form

$$Lu \equiv (D_t^m - A) u = f, \tag{1}$$

where the operation D_t and the operator **A** satisfy the hypotheses of Chapter V. In the simplest case, which is all that we consider, **A** is assumed to be a Π-operator. As before, we preserve the definitions and notation of Chapter V.

As in §1, the study of the operator equation (1) reduces essentially to the study of formulas that yield solutions of the corresponding differential equations (in which **A** is a numerical parameter) for various boundary conditions.

Let us consider the characteristic equation connected with (1):

$$z^m - A = 0. \tag{2}$$

With each complex number A we may associate numbers $m_+(A)$, $m_-(A)$, $m_0(A)$, the numbers of roots of (2) for which $\operatorname{Re} z > 0$, $\operatorname{Re} z < 0$, and $\operatorname{Re} z = 0$.

Without showing the arguments of A ($A = A_s$) explicitly, we observe that there are the following possibilities for the distribution of the roots of (2) in the complex plane:

for even m:

$$\text{I-a} \quad m_+ = m_-, \qquad m_0 = 0; \qquad \text{I-b} \quad m_+ = m_-, \quad m_0 = 2;$$

for odd m: $\hspace{10cm}$ (3)

$$\text{II-a} \quad m_+ = m_- \pm 1, \quad m_0 = 0; \qquad \text{II-b} \quad m_+ = m_-, \quad m_0 = 1.$$

Turning now to the case $A = A(-iD)$, we associate, as usual, with the operation $A(-iD)$ the polynomial $A(s)$, $s \in S$, and the Π-operator A: $\mathbb{H}_x \to \mathbb{H}_x$. Observing that for each value of $A(s)$ there are corresponding values of the numbers $m_+(s)$, $m_-(s)$, $m_0(s)$ in Table (3), we introduce the following definition.

Definition. The operator A: $\mathbb{H}_x \to \mathbb{H}_x$ is *stable* if there is an $N > 0$ such that when $|s| \geq N$ each of the numbers $m_+(s), m_-(s), m_0(s)$ has a constant value. In the contrary case A is *unstable*.

Therefore for a stable operator the numbers m_+, m_-, m_0 are independent of $s \in \mathcal{S}$ for sufficiently large s.

We now consider the simplest ("split") conditions on t of the following form:

$$D_t^k u|_{t=0} = 0, \qquad k = 0, 1, \dots, k_0 - 1; \tag{Γ_0}$$
$$D_t^k u|_{t=b} = 0, \qquad k = 0, 1, \dots, k_b - 1; \tag{Γ_b}$$
$$k_0 + k_b = m, \ 0 \leq k_0, \ k_b \leq m.$$

Here it is assumed that the condition $k_0 = 0$ (or $k_b = 0$) means that there is no condition for the corresponding value of t. If both $k_0 \neq 0$ and $k_b \neq 0$ then the operator D_t^m: $\mathbb{H}_t \to \mathbb{H}_t$ defined by these conditions has a nonempty point spectrum whose structure can be studied without difficulty. This situation is discussed in, for example, [18].

Theorem 1. *Let A be a stable operator, let the numbers m_+, m_-, and m_0 be defined by (3), and let conditions (Γ_0) and (Γ_b) be chosen so that*

$$
\begin{array}{lll}
k_0 = m_- & & \text{in case I-a;} \\
k_0 = m_- & \text{or} \quad k_0 = m_- + 2 & \text{in case I-b;} \\
k_0 = m_- & & \text{in case II-a;} \\
k_0 = m_- & \text{or} \quad k_0 = m_- + 1 & \text{in case II-b.}
\end{array}
\tag{4}
$$

Then the point spectrum $P\sigma L$ of the corresponding operator $\mathbf{L}: \mathbb{H} \to \mathbb{H}$ (if $P\sigma D_t^m \neq \emptyset$) consists of the points

$$\lambda_{p,s} = \lambda_p - \mathbf{A}(s), \qquad \lambda_p \in P\sigma D_t^m, \ s \in \mathscr{S}. \tag{5}$$

The complement of the closure of $P\sigma L$ in $\mathbb{C}$ belongs to the resolvent set of $\mathbf{L}$. If $P\sigma D_t^m = \emptyset$ then $\rho\mathbf{L} = \mathbb{C}$.

If the choice of (split) conditions on t contradicts (4), then, for the corresponding operator $\mathbf{L}$, each point of the complex plane $\mathbb{C}$ belongs either to the point spectrum or the continuous spectrum. Case I-b is exceptional; here the choice $k_0 = m_- + 1$ leads to the existence of an unstable point spectrum. $\square$

We make some additional remarks. The case $P\sigma D_t^m = \emptyset$ in the set of conditions that satisfy the requirements of Table (4) can occur only for the cases $m = 1$ or 2, which have already been discussed in detail.

By instability of the point spectrum we mean a phenomenon analogous to that considered in 1.5 of the preceding section, in the remark on the Dirichlet problem for a hyperbolic operator. Including a detailed description of it in the theorem would cause too much complication.

If we omit the requirement of the stability of the operator $\mathbf{A}$ we can evidently again encounter operators whose spectra fill the complex plane. It is easy to see, however, that even in this case the construction used for equation (1), described in 1.3 of the preceding section, enables us to define a proper operator.

To describe proper operators $\mathbf{L}: \mathbb{H} \to \mathbb{H}$ by using standard boundary conditions for operators $\mathbf{A}$ that are "not too bad", it turns out to be necessary (as in the cases $m = 1, 2$ that we studied in detail) to extend the class of boundary conditions by adjoining to the splitting conditions (Γ_0) and (Γ_b) the nonlocal conditions

$$\mu_1 D_t^{k_1} u|_{t=0} - D_t^{k_1} u|_{t=b} = 0,$$
$$\mu_2 D_t^{k_2} u|_{t=0} - D_t^{k_2} u|_{t=b} = 0.$$

In speaking of the adjoined conditions (Γ_μ) we are always to understand that $\mu \neq 0$ or ∞.

Theorem 2. *Let the operator A be unstable. Then if*

a) $m = 2q + 1 \geq 3$ *is odd and one condition (Γ_μ), $k_1 = q + 1$, is adjoined to the boundary conditions $\Gamma_0, \Gamma_b, k_0 = k_b = q$, or*

b) $m = 2q \geq 4$ *is even and two conditions of the form (Γ_μ), $k_1 = q$, $k_2 = q + 1$, are adjoined to the boundary conditions $\Gamma_0, \Gamma_b, k_0 = k_b = q - 1$,*

then the point spectrum of the corresponding operator $\mathbf{L}: \mathbb{H} \to \mathbb{H}$ consists of points of the form (5), and the complement of the closure of $P\sigma L$ in $\mathbb{C}$ belongs to the resolvent set of $\mathbf{L}$.

If, however, the boundary conditions on t are taken to be splitting (contain only conditions of the forms Γ_0 and Γ_b) then the complement of the point spectrum (which has the structure described in (5)) belongs to the continuous spectrum of **L.** □

Remark. It does not follow from the conclusion of the theorem that in case b) it is impossible to get along with one nonlocal condition.

2.2. The General Operator Equation. In turning to the general equation of the form (1), §0, we give up, as remarked previously, the possibility of writing general formulas for the dependence of the roots of the characteristic equation

$$\sum_j \mathbf{A}_{m-j} z^j = 0 \qquad (6)$$

on the coefficients $\mathbf{A}_k$ (such formulas are still available for binomial equations, i.e. for the roots of equation (2)). Therefore, supposing that the operators $\mathbf{A}_k$ are determined by some distribution of the roots of (6), this necessarily has an implicit form. If we postulate a particular kind of distribution of the roots of (6) for all possible values of the parameters $\mathbf{A}_k = \mathbf{A}_k(s)$, $s \in \mathscr{S}$, we can single out different classes of operations and appropriate standard boundary conditions that define corresponding regular operators. As in the discussion of binomial equations, the choice of "appropriate" standard conditions is determined by the set of roots of (6) that satisfy the requirements $\operatorname{Re} z \gtrless 0$, $\operatorname{Re} z = 0$, and the type of stability that fulfills these requirements [40].

Finally, it is possible to find special subclasses of equations that can be studied more effectively (as discussed in detail in subsection 1 for binomial equations).

For example, we may suppose that the operation $L(D)$ is given in factored form, i.e. the chain of ordinary differential equations by means of which we determine the coefficients $u_s(t)$ of the desired solution has the form

$$(D_t - \mathbf{B}_1(s))(D_t - \mathbf{B}_2(s))\ldots(D_t - \mathbf{B}_m(s)) u_s = f_s, \qquad s \in \mathscr{S}, \qquad (7)$$

where $\mathbf{B}_k(s)$ are polynomials determined by the operations $\mathbf{B}_k(-iD)$ (or by Π-operators $\mathbf{B}: \mathbb{H}_x \to \mathbb{H}_x$).

Representing (7) in the form

$$(D_t - \mathbf{B}_1(s)) \mathbf{L}_{m-1,s} u_s = f_s, \qquad (8)$$

we suppose that the condition

$$\mu_{m-1} \mathbf{L}_{m-1,s} u_s|_{t=0} - \mathbf{L}_{m-1,s} u_s|_{t=b} = 0 \qquad (9)$$

ensures, for some μ_{m-1} (possibly $\mu_{m-1} = \mu_{m-1}(s)$), the solvability of (8) with respect to $\mathbf{L}_{m-1,s} u_s$, and that by continuing this process we can find u_s.

When corresponding uniform inequalities for the norm of $u_s(t)$ are satisfied, conditions of the form (9) will determine a proper operator.

In many cases it turns out to be possible to substitute for (9) the chain of conditions

$$\mu D_t^k u_s|_{t=0} - D_t^k u_s|_{t=b} = 0, \qquad k = 0, 1, \ldots, m-1, \tag{10}$$

i.e., to suppose that the operations D_t^j in $\mathbf{L}(D)$ generate powers of the operator D_t determined by the conditions

$$\mu u|_{t=0} - u|_{t=b} = 0. \tag{11}$$

If, for example, $\mu_1 = \ldots = \mu_{m-1} = \mu$ independently of s in conditions of the form (9), then condition (11) let us reduce the condition

$$\mu(D_t - \mathbf{B}_m(s)) u_s|_{t=0} - (D_t - \mathbf{B}_m(s)) u_s|_{t=b} = 0$$

to the form

$$\mu D_t u_s|_{t=0} - D_t u_s|_{t=b} = 0,$$

etc.

Turning to the general case, we observe that under the additional hypothesis $\mathbf{A}_0 \equiv 1$ the construction of 1.4, §1, evidently lets us describe the proper operator $\mathbf{L}$ for arbitrary $\varPi$-operators $\mathbf{A}_k$, $k = 1, \ldots, m$.

To remove the requirement that $\mathbf{A}_0 \equiv 1$, i.e. to describe a class of regular operators associated with an arbitrary differential operation with constant coefficients, requires further discussion, which will be given in the next chapter.

Chapter VII
General Existence Theorems for Proper Operators

§0. Introductory Remarks

This chapter contains three sections. In the first section we establish the fact (sufficiently obvious by itself) that a proper operator defined in a domain $V \subset \mathbb{R}^n$ (satisfying our usual requirements) automatically induces a proper operator in a domain $V' \subset V$.

By using this property we can establish the existence of a proper operator generated by a given differential operation with constant coefficients, in any bounded domain V. It is sufficient to locate this region in a parallelepiped of sufficiently large size, bounded by planes parallel to the coordinate axes. In such a parallelepiped, on the assumption that the coordinate axes are not characteristic directions (which we can always bring about by a rotation of the axes) the proper operator is given by the construction of 1.4 in the preceding chapter. This also proves Hörmander's theorem which we mentioned in 3.3, Chapter II. This is the content of §2.

The last section discusses the problem of the description of a proper operator generated by an arbitrary operation with constant coefficients in a given parallelepiped bounded by planes parallel to the coordinate axes, without the hypothesis that the axes are not characteristic directions. This requires an essential modification of the construction, but on the other hand provides an effectivization (and somewhat sharper form) of Hörmander's theorem for this class of domains.

§1. Lemma on Restriction of a Domain

A domain $V \subset \mathbb{R}^n$ that satisfies our usual conditions (§1, Chapter II) will be said to be *admissible*.

Let $V' \subset V \subset \mathbb{R}^n$ be admissible domains, $\mathbb{H} \equiv \mathbb{H}(V)$, $\mathbb{H}' \equiv \mathbb{H}(V')$ the corresponding Hilbert spaces and for some differential operation $L(D)$ let the proper operator $L : \mathbb{H} \to \mathbb{H}$ be defined in V.

Lemma. *Under the preceding hypotheses we can always associate with L an operator $L' : \mathbb{H}' \to \mathbb{H}'$ which is proper and generated by $L(D)$ on V'.*

Proof. We shall construct an operator $\mathbf{L}'$ with the required properties. Let $\mathbb{H}_0$ denote the subspace of $\mathbb{H}$ formed by the elements that are identically zero outside V'. We shall consider an element $v \in \mathbb{H}'$ to belong to $\mathfrak{D}(\mathbf{L}')$ if there is an element $u \in \mathfrak{D}(\mathbf{L})$ such that $\mathbf{L}u \in \mathbb{H}_0$, $u|_{V'} = v$. Set $\mathbf{L}'v = \mathbf{L}u$.

The operator so constructed is an extension of the minimal operator $\mathbf{L}'_0$ generated by $\mathbf{L}(D)$ on V'. In fact, if $v \in \mathfrak{D}(\mathbf{L}'_0)$ there is a sequence $\{v_i\} \in C_0^\infty(V')$ such that $v_i \to v$ in H' and $\mathbf{L}(D)v_i = f_i \to f$. If we continue v_i to be identically zero outside V', we obtain a sequence $\{\tilde{v}_i\} \in C_0^\infty(V)$, which lets us establish that the limit function $\tilde{v} \in \mathbb{H}$ belongs to $\mathfrak{D}(L_0) \subset \mathfrak{D}(L)$, and

$$\tilde{v}|_{V'} = v, \quad \mathbf{L}\tilde{v} \in \mathbb{H}_0.$$

We can show similarly that the operator so constructed is a restriction of the maximal operator $\tilde{\mathbf{L}}'$.

The equation $\mathbf{L}'v = f$ has a unique solution for every right-hand side $f \in \mathbb{H}'$. This follows immediately from the uniqueness of the element $u \in \mathfrak{D}(\mathbf{L})$ such that $\mathbf{L}u = \tilde{f}$, where $\tilde{f} \in \mathbb{H}$ is obtained by continuing f to be zero outside V'. $\square$

Example. For $n = 1$, for the differential operation

$$\mathbf{L}(D) \equiv D_t - a, \quad a = \text{const},$$

and under the hypothesis that the proper operator $\mathbf{L}: \mathbb{H} \to \mathbb{H}$ is determined by the conditions

$$\mu u|_{t=0} - u|_{t=b} = 0,$$

it is easy to give an explicit description of the boundary conditions determining the operator $\mathbf{L}'$ described above. It is enough to substitute into the formula for the solution of the regular problem for the equation $\mathbf{L}u = f$ an element f that is identically zero outside $V' = (b_1 < t < b_2)$, where $0 < b_1 < b_2 < b$. For $v \in \mathfrak{D}(L')$ we obtain the equation

$$\mu e^{b_2 a} v|_{t=b_1} - e^{(b+b_1)a} v|_{t=b_2} = 0. \tag{1}$$

Here the condition $\mu - e^{ab} \neq 0$ automatically guarantees the regularity of the problem corresponding to (1).

In conclusion, it is useful to observe that if we replace $\mathbb{H}_0$ by $\mathbb{H}_\varphi$ in the construction used in the lemma, where $\mathbb{H}_\varphi$ is the subspace of $\mathbb{H}$ composed of the elements that coincide with a given element φ on $V \backslash V'$, $\varphi \not\equiv 0$, then the operator $\mathbf{L}'_\varphi$ so defined will not, in general, yield an extension of $\mathbf{L}'_0$.

§2. Existence Theorem for a Proper Operator

In $\mathbb{R}^n$ let there be given a general differential operation $\mathbf{L}(D)$ of order m with constant complex coefficients. If we select one of the coordinates

$x_1, \ldots, x_n$, we may write $L(D)$ in the form

$$L(D) \equiv A_0 D^k + A_1 D^{k-1} + \ldots + A_k,$$

where D is the operation of differentiation with respected to the selected coordinate, and the operations $A_0, \ldots, A_k$ involve differentiation only with respect to the other $n-1$ variables. If the selected coordinate can be chosen so that $L(D)$, possibly after division by a constant, can be written in the form

$$L(D) \equiv D^k + A_1 D^{k-1} + \ldots + A_k, \tag{1}$$

then the differential operation (written in the form (1)) will be said to be *reduced*.

We shall suppose that the selected coordinate is x_1. We choose an integer $N > 0$ and call a parallelepiped Q of the form

$$Q: (0 < x_1 < b) \times (0 < x_2 < 2N\pi) \times \ldots \times (0 < x_n < 2N\pi)$$

a *standard parallelepiped* in $\mathbb{R}^n$.

Lemma 1. *In a standard parallelepiped, for every differential operation $L(D)$, as described above, with constant coefficients, there is a proper operator generated by it.*

Proof. It is evidently enough to consider the case $N = 1$, using the technique of the proof of Theorem 2, 1.4, Chapter VI.

In the first place, there is a constant $M > 0$ such that for arbitrary complex numbers $A_1, \ldots, A_k$ a proper operator $L: \mathbb{H}_1 \to \mathbb{H}_1$ (where $\mathbb{H}_1$ is the Hilbert space on the interval $0 < x_1 < b$), generated by the operation (1) $(D \equiv D_1)$ can be defined (with corresponding conditions for $x_1 = 0, b$) so that $\|L^{-1}\| \leq M$. The proof of this statement is a repetition of the proof of the lemma in 1.4, Chapter VI.

Moreover, if we take the boundary conditions on $x_2, \ldots, x_n$ to be the periodicity conditions (i.e. if we define $A_1, \ldots, A_k$ as the corresponding Π-operators), use our general method of decomposing the equation

$$L(D)u = f$$

into a chain of ordinary differential equations

$$L_s(D_1)u_s = f_s, \quad s \in \mathcal{S},$$

and repeat the argument for the proof of Theorem 2, §1, Chapter VI, we obtain what was required. $\square$

From Lemma 1 and the lemma in the preceding section we immediately obtain the following corollary.

Corollary. *If* **L**(D) *is a reduced differential operation with constant complex coefficients and* V *is an admissible domain in* $\mathbb{R}^n$, *there is always a proper operator*

$$\mathbf{L}\colon\ \mathbb{H}(V)\to\mathbb{H}(V),\tag{2}$$

generated by **L**(D).

Proof. We may suppose without loss of generality that V is contained in $\mathbb{R}^n_+$ (i.e. in the part of $\mathbb{R}^n$ where $x_k>0$, $k=1,\dots,n$) and that Q is a standard parallelepiped such that $V\subset Q$. Then the proper operator **L**: $\mathbb{H}(Q)\to\mathbb{H}(Q)$ generated by **L**(D), whose existence is guaranteed by Lemma 1, determines, according to the lemma in § 1, a proper operator (2). $\square$

Now, in order to turn to the case of an arbitrary (not reduced) operation **L**(D) with constant coefficients, we consider the question of how the entities under investigation behave under smooth invertible changes of coordinates in $\mathbb{R}^n$, or equivalently under smooth one-to-one mappings of $\mathbb{R}^n$ on itself.

Consider a transformation

$$\varphi\colon\ \mathbb{R}^n\to\mathbb{R}^n$$

given by the formula

$$x=\varphi(x').\tag{3}$$

The operation **L**(D) defined in the coordinates $\{x\}$ is transformed by (3) into an operation **L**'(D') in the coordinates $\{x'\}$. The form of **L**'(D') is determined by the classical formulas for change of variables.

Lemma 2. *Let* **L**(D) *be an unreduced differential operation with constant coefficients. Then there is a coordinate transformation* (3), *in fact a rotation, such that in the new coordinates the transformed operation is reduced.*

Proof. We consider the group of leading terms in **L**(D):

$$\mathbf{L}_m(D)=\sum_{|\alpha|=m}a_\alpha D^\alpha.$$

Let the required coordinate transformation be given by

$$x'_k=\sum_j \gamma^j_k x_j,\qquad k=1,\dots,n,$$

where $\{\gamma^j_k\}$ is an orthogonal matrix. Calculating the coefficients of, for example, $(D'_1)^m$ in the new variables, we will have

$$D^{\alpha_1}_1\dots D^{\alpha_n}_n u=(\gamma^1_1)^{\alpha_1}\dots\dots(\gamma^1_n)^{\alpha_n}(D'_1)^m u'+\mathbf{L}_{\alpha,m}(D')u',$$

where $\mathbf{L}_{\alpha,m}(D')$ does not contain the derivatives $(D'_1)^m$. Correspondingly

$$\mathbf{L}'_m(D')=\sum_{|\alpha|=m}\{a_\alpha\gamma^{1,\alpha}(D'_1)^m+\mathbf{L}_{\alpha,m}(D')\}.$$

There are evidently infinitely many ways of choosing an orthogonal matrix $\{\gamma_k^j\}$ for which

$$\sum_{|\alpha|=m} a_\alpha \gamma^{1,\alpha} \neq 0. \quad \square$$

Now let V' be the pre-image of the admissible domain V under the transformation

$$\varphi: V' \to V \tag{4}$$

generated by (3). Let $\mathbb{H}' \equiv \mathbb{H}(V')$, $\mathbb{H} \equiv \mathbb{H}(V)$. Consider the correspondence between the proper operators

$$\mathbf{L}: \mathbb{H} \to \mathbb{H}, \qquad \mathbf{L}': \mathbb{H}' \to \mathbb{H}'$$

(generated by $\mathbf{L}(D)$ and $\mathbf{L}'(D')$) induced by (4).

The transformation (4) generates, in the usual way, a transformation

$$\varphi^*: \mathbb{H} \to \mathbb{H}'.$$

To determine it, it is enough to consider the corresponding transformation

$$\varphi^*: u(x) \to u'(x')$$

for functions that are elements of $C(V)$, and use the fact that C is dense in $\mathbb{H}$.

Now let $\mathbf{L}': \mathbb{H}' \to \mathbb{H}'$ be a proper operator generated by $\mathbf{L}'(D')$. For $u \subset \mathbb{H}$, set

$$\mathbf{L}u = \varphi^{*-1} \mathbf{L}' \varphi^* u, \tag{5}$$

assuming that $u \in \mathfrak{D}(\mathbf{L})$ if and only if $\varphi^* u \in \mathfrak{D}(\mathbf{L}')$. The equation

$$\mathbf{L}u = f$$

then evidently has a unique solution for every element $f \in \mathbb{H}$.

We observe in addition that if $\mathbf{L}_0'$ and $\tilde{\mathbf{L}}'$ are the minimal and maximal operators generated on $\mathbb{H}'$ by $\mathbf{L}'(D')$, then, since $\mathbf{L}(D)$ and $\mathbf{L}'(D')$ are connected, as above, by the classical formulas, and if $\mathbf{L}_0, \tilde{\mathbf{L}}, \mathbf{L}_0'$, and $\tilde{\mathbf{L}}'$ are defined as the closures of the corresponding operations on smooth functions, it follows immediately from (3) and (5) that

$$\mathbf{L}_0 \subset \mathbf{L} \subset \tilde{\mathbf{L}},$$

i.e. $\mathbf{L}$ is a proper operator generated by $\mathbf{L}(D)$.

We state this result as a lemma.

Lemma 3. *Let V be an admissible domain, V' the pre-image of V under the transformation* (3)*, and $\mathbf{L}': \mathbb{H}' \to \mathbb{H}'$ a proper operator generated by $\mathbf{L}'(D')$.*

Then the operator L: $\mathbb{H} \to \mathbb{H}$ *defined by* (5) *is a proper operator generated by* L(D).

We state the final result of this section as a theorem.

Theorem. *If* L(D) *is any differential operation with constant coefficients and V is an admissible domain in* $\mathbb{R}^n$, *then there is always a proper operator*

$$L: \mathbb{H}(V) \to \mathbb{H}(V),$$

generated by L(D).

Proof. If L(D) is reduced, the required result is given by the corollary of Lemma 1, above.

If L(D) is an unreduced differential operation then (using the notation introduced above and Lemma 2) we can say that there is a coordinate transformation (a rotation) that makes L(D) correspond to a reduced operation L′($D′$), again with constant complex coefficients Then there is an operator

$$L': \mathbb{H}' \to \mathbb{H}'$$

generated by L′($D′$) on $\mathbb{H}'$ (corollary of Lemma 1) and the operator

$$L: \mathbb{H} \to \mathbb{H}$$

defined by (5) provides, by Lemma 3, a proper operator generated by L(D) on $\mathbb{H}(V)$. □

§3. Description of Proper Operators in a Parallelepiped

3.0. Preparatory Remarks. This section discusses the problem of a direct description, for any operator L(D), of a proper operator that it generates, in a given parallelepiped Q, i.e. a description that does not exclude the case when L(D) is not reduced.

The existence in this case of a proper operator L: $\mathbb{H}(Q) \to \mathbb{H}(Q)$ generated by L(D) is a corollary of the theorem in the preceding section. However, the proof of that theorem does not provide a sufficiently explicit description of the domain of L. In the case when L(D) is a reduced operation, the description is given by Lemma 1, §2. Now we need to find a similar description for a general L(D) (in the parallelepiped Q).

The construction that we use, suggested in [36], is of independent interest. It makes essential use of the form of the region (a parallelepiped) that we consider. The possibility of some effective realization of the general theorem of the preceding section for a general $V \subset \mathbb{R}^n$ remains open.

3.1. Description of a Proper Operator by an Appropriate Choice of Basis. As one can easily predict (and as will follow from the discussion below), in order to construct, in any parallelepiped, a proper operator generated by a general operation $\mathbf{L}(D)$ with constant coefficients, it is enough to be able to construct the operator in the standard cube

$$V = \prod_1^n (0 < x_i < 2\pi).$$

Thus, let $\mathbf{L}(D)$ be any given operation with constant coefficients, and V this cube. We define a proper operator $\mathbf{L}: \mathbb{H}(V) \to \mathbb{H}(V)$ generated by $\mathbf{L}(D)$ by using the following procedure.

To a real number $\alpha_1, 0 \leq \alpha_1 < 1$, we assign a Riesz basis on the interval $l_1 = [0 \leq x_1 \leq 2\pi]$, consisting of exponentials

$$\{e^{i(\alpha_1 + k)x_1}\} \qquad k = 0, \pm 1, \pm 2, \dots. \tag{1}$$

Then each element $u \in \mathbb{H}(V)$ can be represented in the form

$$u(x) = \sum_k u_k(x_2, \dots, x_n) e^{i(\alpha_1 + k)x_1}. \tag{2}$$

Substituting the formal representation (2) for u and f into the equation

$$\mathbf{L}(D) u = f \tag{3}$$

and equating the coefficients of $e^{i(\alpha_1 + k)x_1}$, we decompose (3) into a chain of equations

$$\mathbf{L}_k(D) u_k = f_k, \qquad k = 0, \pm 1, \pm 2, \dots, \tag{4}$$

where the operations $\mathbf{L}_k(D)$ contain differentiations only with respect to $x_2, \dots, x_n$.

Let us suppose that $\mathbf{L}_k(D) \not\equiv 0$ for each k (as will be shown below, this can always be arranged by a suitable choice of α_1), and associate with each $\mathbf{L}_k(D)$, in some way (supposing that this is possible) a proper operator

$$\mathbf{L}_k: \ \mathbb{H}_1(V_1) \to \mathbb{H}_1(V_1),$$

where V_1 is the cube with edge 2π in $(x_2, \dots, x_n)$ space. In addition, we suppose that the norms of the operators $\mathbf{L}_k^{-1}$ are bounded, uniformly in k:

$$\|\mathbf{L}_k^{-1}\| \leq c \tag{5}$$

(see Lemmas 3–5, below). We now define the operator $\mathbf{L}: \mathbb{H}(V) \to \mathbb{H}(V)$ by defining it on finite sums of the form (2), subject to the additional requirement $u_k \in \mathfrak{D}(L_k)$, by the equation

$$L u = \sum_{|k| \le N} L_k u_k e^{i(\alpha_1 + k)x_1},$$

and taking its closure in $\mathbb{H}$.

Lemma 1. *Under the preceding hypotheses, the operator* $L: \mathbb{H} \to \mathbb{H}$ *described above is a proper operator generated by the operation* $L(D)$.

Proof. It is evident that the equation $Lu = f$ has a unique solution for each right-hand side $f \in \mathbb{H}$. (The solution is given by the series (2), with $u_k = L_k^{-1} f_k$.) It is clear that $L \subset \tilde{L}$, i.e. L is a restriction of the maximal operator generated by $L(D)$.

To verify that $L_0 \subset L$, i.e. that L is an extension of the minimal operator, we observe that the basis (1) consists of the eigenfunctions of the operator $D_1: \mathbb{H}(l_1) \to \mathbb{H}(l_1)$, $l_1 = (0 < x_1 < 2\pi)$ determined by the conditions

$$e^{i\alpha_1} u|_{x_1 = 0} - u|_{x_1 = 2\pi} = 0. \tag{Γ_1}$$

Hence it follows that the smooth finitely-supported functions belong to the domain of every power of the operator D_1. If now $u \in \mathfrak{D}(L_0)$, and $\{u_i\}$ is an approximating sequence of smooth finitely-supported functions ($u_i \to u$, $L(D) u_i \to g$, with convergence in $\mathbb{H}$), then each u_i is represented by a series of the form (2) that admits termwise differentiation (of every order) with respect to x_1 and termwise application of the operators L_k (since the functions $u_k(x_2, \ldots, x_n)$ are smooth and finitely supported, they automatically belong to $\mathfrak{D}(L_k)$). Consequently every function $u \in \mathfrak{D}(L_0)$ is representable by the series (2) and provides a solution (in the sense described above) of the equation $Lu = f$. $\square$

We now observe that the construction just described is in turn applicable to each operation $L_k(D)$. Therefore if we choose a real number $\alpha_2(k)$, $0 \le \alpha_2(k) < 1$, and choose, on the interval $l_2 = [0 \le x_2 \le 2\pi]$, the basis

$$\{e^{i[\alpha_2(k) + s]x_2}\}, \qquad s = 0, \pm 1, \pm 2, \ldots,$$

and make the corresponding assumptions about the operations $L_{ks}(D)$ (obtained by decomposing each of the operations (4)), we can define a proper operator $L: \mathbb{H} \to \mathbb{H}$ as the closure in $\mathbb{H}$ of the operator defined on the corresponding finite sums by the equation

$$L u = \sum_{|k| \le N, |s| \le M} L_{ks} u_{ks} e^{i[(\alpha_1 + k)x_1 + (\alpha_2(k) + s)x_2]}.$$

At the next step the construction can be applied to the operations $L_{ks}(D)$, etc.

The final result obtained after n steps can be stated in two lemmas.

Lemma 2. *The system of exponentials $\varepsilon(s)$, $s \in \mathcal{S}$, of the form*

$$\varepsilon(s) \equiv \varepsilon(s_1, \dots, s_n) = \exp i \{ [\alpha_1 + s_1] x_1$$
$$+ [\alpha_2(s_1) + s_2] x_2 + \dots + [\alpha_n(s_1, \dots, s_{n-1}) + s_n] x_n \}, \qquad (6)$$

where the values of the functions α_k, $k = 1, \dots, n$, lie in the half-open interval $[0, 1)$, form a Riesz basis in $\mathbf{H}(V)$. □

Lemma 3. *If the basis (6) has the property that, for the given operation $L(D)$ with constant coefficients, the inequality*

$$|L(D) \varepsilon(s)| \geq \delta > 0 \qquad (7)$$

is satisfied for every $s \in \mathcal{S}$, then the operator $L: \mathbf{H} \to \mathbf{H}$, defined as the closure in $\mathbf{H}$ of the operator defined on finite sums $u = \sum_s u_s \varepsilon(s)$, $|s_k| \leq N_k$, $k = 1, \dots, N$, by the equation

$$Lu = \sum_s u_s L(D) \varepsilon(s),$$

is a proper operator generated by $L(D)$.

Proof. It is immediately obvious that the operator L described in the hypothesis of the lemma is a restriction of the maximal operator $\tilde{L}$, and that the equation $Lu = f$ has a unique solution for every $f \in \mathbf{H}$. It remains to verify that $L_0 \subset L$, i.e. that L is an extension of the minimal operator. However, this follows from the identity of the operator L of our lemma with the proper operator determined by the process above, where in the last step we used the operator $L_{s_1, \dots, s_{n-1}}$ associated with the equation

$$L_{s_1, \dots, s_{n-1}}(D) u_{s_1, \dots, s_{n-1}}(x_n) = f_{s_1, \dots, s_{n-1}}(x_n),$$

which is solved under the boundary conditions

$$e^{i \alpha_n(s_1, \dots, s_{n-1})} u_{s_1, \dots, s_{n-1}}|_{x_n = 0} - u_{s_1, \dots, s_{n-1}}|_{x_n = 2\pi} = 0,$$

i.e. by using the expansions of functions of x_n in terms of the basis

$$\{ \exp i [\alpha_n(s_1, \dots, s_{n-1}) + s_n x_n] \}, \qquad s_n = 0, \pm 1, \pm 2, \dots.$$

Condition (7) guarantees the feasibility of all n steps. □

3.2. Existence of a Correctly Chosen Basis. Now in order to be able, for an arbitrary operation $L(D)$ with constant coefficients, to specify a proper operator generated by it in the cube V, it remains to verify that for every such operation there is a basis of the form (6) that satisfies (7).

Lemma 4. *Let*

$$P(z) \equiv a_n z^n + \ldots + a_0 \tag{8}$$

be a given polynomial with constant complex coefficients. Then there is a real number α, $0 \leq \alpha < 1$, *such that the inequality*

$$|P[i(\alpha + k)]| \geq |a_n|(2n)^{-n}$$

is satisfied for all $k = 0, \pm 1, \pm 2, \ldots$.

Proof. Let T be the circle of unit length obtained by identifying the endpoints of the interval $[0,1]$, with a corresponding parametrization and the natural metric $\rho(t_1, t_2) = \min(|t_1 - t_2|, |t_1 - t_2 \pm 1|)$. Let $F \colon \mathbb{C} \to T$ map the complex plane on this circle, making the point $z = z_1 + i z_2$ correspond to the point $F(z) = \{z_2\}$ (the fractional part of the real number z_2). Evidently $|z' - z''| \geq \rho(F(z'), F(z''))$. Let $\xi_1, \ldots, \xi_n$ be the zeros of the polynomial (8). Then there is a point $\alpha \in T$ such that $\rho(\alpha, F(\xi_p)) \geq 2^{-n}$ for $p = 1, \ldots, n$. This in turn implies the chain of inequalities

$$|P[i(\alpha + k)]| = |a_n| \prod_{p=1}^{n} |[i(\alpha + k) - \xi_p]|$$

$$\geq |a_n| \prod_{p=1}^{n} \rho(\alpha, F(\xi_p)) \geq |a_n|(2n)^{-n}. \quad \square$$

Lemma 5. *For every given operation* $\mathbf{L}(D)$ *with constant coefficients there is a basis of the form* (6) *such that* (7) *holds.*

Proof. Let $\mathbf{L}(\xi)$ be the complex polynomial associated in the usual way with $\mathbf{L}(D)$:

$$\mathbf{L}(D) e^{ix \cdot \xi} = \mathbf{L}(\xi) e^{ix \cdot \xi}.$$

We represent $\mathbf{L}(\xi)$ in the form

$$\mathbf{L}(\xi) = \mathbf{L}_1(\xi_1, \ldots, \xi_{n-1})(\xi_n^{m_1} + q_n^1 \xi_n^{m_1 - 1} + \ldots + q_n^{m_1}) \equiv \mathbf{L}_1(\xi_1, \ldots, \xi_{n-1}) Q_1(\xi_1, \ldots, \xi_n).$$

Here m_1 is the degree of $\mathbf{L}(\xi)$ in ξ_n, $\mathbf{L}_1$ is the polynomial coefficient of $\xi_n^{m_1}$, and q_n^p, $p = 1, \ldots, m_1$, is a uniquely defined rational function of $\xi_1, \ldots, \xi_{n-1}$. If $m_1 = 0$ we have $\mathbf{L}_1 \equiv \mathbf{L}$. Continuing this process, we obtain

$$\mathbf{L}(\xi) = \mathbf{L}_1 Q_1 = \mathbf{L}_2(\xi_1, \ldots, \xi_{n-2}) Q_2(\xi_1, \ldots, \xi_{n-1}) Q_1(\xi_1, \ldots, \xi_n) = \ldots$$
$$= \mathbf{L}_{n-1}(\xi_1) Q_{n-1}(\xi_1, \xi_2) \ldots Q_1(\xi_1 \ldots \xi_n).$$

Let $M = \max m_p$, $p = 1, \ldots, n$. By using Lemma 4 we can find a number $\alpha_1, 0 \leq \alpha_1 < 1$, such that

$$|\mathbf{L}_{n-1}[i(\alpha_1+s_1)]|\geq |a|(2M)^{-M}.$$

Now to each s_1 there corresponds a certain polynomial

$$Q_{n-1,s_1}(\xi_2)\equiv Q_{n-1}[i(\alpha_1+s_1),\xi_2],$$

for which we can again, by Lemma 4, choose a number $\alpha_2=\alpha_2(s_1)$, $0\leq\alpha_2<1$, such that

$$|Q_{n-1,s_1}[i(\alpha_2(s_1)+s_2)]|\geq(2M)^{-M}$$

for all integers s_2. At the next step we choose in the same way a number $\alpha_3(s_1,s_2)$ for the polynomials

$$Q_{n-2,s_1,s_2}(\xi_3)$$

and so on.

It is now clear that for a basis of the form (6) in which the values of α_k are chosen in the indicated way, the inequality

$$|\mathbf{L}(D)\varepsilon(s)|\geq |a|(2M)^{-Mn}$$

will be satisfied for all $s\in\mathscr{S}$. $\square$

3.3. The Final Result. The preceding discussion can be summarized in the following theorem.

Theorem. *In the standard cube V, we can associate with every given operation* $\mathbf{L}(D)$ *with constant complex coefficients a proper operator* $\mathbf{L}$: $\mathbb{H}(V)\to\mathbb{H}(V)$, *defined as the closure in* $\mathbb{H}(V)$ *of the operation* $\mathbf{L}(D)$ *given on finite sums of the form* $\sum_s u_s\varepsilon(s)$, *where* $\{\varepsilon(s)\}_{s\in\mathscr{S}}$ *is a special (correctly chosen) Riesz basis (6) constructed from the polynomial* $\mathbf{L}(\xi)$. $\square$

To pass to the case of an arbitrary parallelepiped, it is sufficient to observe that a nondegenerate parallelepiped can be mapped on the standard cube by an invertible change of variables of the form

$$x'_k=\beta_k x_k, \qquad \beta_k=\text{const}\neq 0, \qquad k=1,\ldots,n.$$

The connection between the special basis constructed for the description of the proper operator and the boundary conditions (used in proving Lemmas 1 and 3) shows that we may pass to a description of $\mathbf{L}$ by means of a chain of recursively defined "special boundary conditions" which are modifications of the special boundary conditions used in §2, Chapter V, or in 1.4 of Chapter VI.

This connection of a correct basis with the boundary conditions also allows us to state a corollary of the theorem which yields another sharpening of Hörmander's results.

Corollary. *If the operation* $L(D)$ *is formally selfadjoint* $(L^t(D) = L(D))$ *then the proper operator* $L: \mathbb{H} \to \mathbb{H}$ *constructed above is selfadjoint.*

In fact, every operator generated by L under conditions of the form (Γ_1) $(\alpha_1$ real$)$ is selfadjoint. This implies the selfadjointness of L. ☐

Chapter VIII
A Special Operational Calculus

§0. Introductory Remarks

As we have already observed, our approach to the study of boundary value problems for partial differential equations by means of reducing them to differential-operator equations is of course also applicable to cases when the operators A_k that appear in equation (1), §0, Chapter VI, are neither Π-operators nor M-operators.

This chapter is devoted to a description of a method that makes it possible to generalize the results presented above to the case when the operators A_k satisfy significantly less restrictive conditions. The method (proposed in [25] and based on [42]) is essentially a development and sharpening of general considerations related to the construction of the operational calculus discussed in 2.2, Chapter I. Naturally the hypotheses about the properties of the operators A_k that are needed in order to obtain interesting results about the solvability of equations of the type (1), §0, Chapter VI, will be stated in terms of the properties of the resolvent.

As we would expect, widening the class of operators A_k makes the constructions less transparent and the results less precise. However, our discussion in any case makes it possible to establish without difficulty that for the differential operation

$$D_t + aD_x, \quad a > 0,$$

considered in the rectangle $V = [0 < t < b] \times [0 < x < 2\pi]$ the Cauchy conditions for $t = 0$ determine a proper operator not only for conditions that are regular in x but also for Cauchy conditions at $x = 0$.

The first section gives the method for constructing the operational calculus in which we are interested, and the second applies the calculus to the study of the properties of the simplest operator equation.

It must be acknowledged that an adequate command of the technique of the operational calculus described in §1 requires the reader to know somewhat more than is outlined in 2.2, Chapter I, about the classical theory of operator-valued functions.

Turning to a description of the contents of the next section, we should note that the previous investigation of operator equations, based on the use of the resolvents of the operators A_k, is poorly suited to obtaining "negative" results, which are an essential part of our discussion. By negative results we mean proofs that whenever the boundary conditions (for the class under consideration) are taken "improperly", or when (for given boundary conditions) the spectrum of A_k does not satisfy the corresponding restrictions, the resulting operator $L\colon \mathrm{IH} \to \mathrm{IH}$ will not be proper.

The proof (in some very simple cases) of the necessity of the restrictions imposed on A_k is discussed in § 3.

The interested reader should compare the technique of this chapter with that proposed in [28].

§ 1. Construction of the Operational Calculus

Although, as before, we shall apply the operational calculus to be described below to our standard space $\mathrm{IH}(V)$, it is natural to base the construction on a Banach space $\mathscr{B}$, since it will not use the additional hypothesis "Hilbert".

Thus let $\mathscr{B}$ be a complex Banach space (B-space) and $\mathbf{T}\colon \mathscr{B} \to \mathscr{B}$ a closed linear operator, generally speaking unbounded, with domain $\mathfrak{D}(\mathbf{T})$ dense in $\mathscr{B}$ and a nonempty resolvent set. Our problem is to construct the broadest possible operational calculus, i.e. to introduce a system of definitions that let us make sense of writing $\varphi(\mathbf{T})$ for the broadest possible class of functions $\varphi(z)$, $z \in \mathbb{C}$, under the fewest possible restrictive requirements on the operator $\mathbf{T}$.

We first construct a linear space $\mathfrak{B}$ such that $\mathscr{B} \subset \mathfrak{B}$ and every power of $\mathbf{T}$ (appropriately extended) is defined on $\mathfrak{B}$. Since we assumed that $\rho \mathbf{T}$ is not empty, we may, without loss of generality, suppose that zero is a regular point, i.e. that $\mathbf{T}^{-1}$ is a bounded operator defined on $\mathscr{B}$.

Let $\mathscr{B}_k$ be the set of pairs (v, k), where $v \in \mathscr{B}$ and k is an integer (positive or negative). We define a linear operation in $\mathscr{B}$ by taking

$$\alpha(v, k) + \beta(w, k) = (\alpha v + \beta w, k)$$

for α and $\beta \in \mathbb{C}$, and introduce the norm

$$\|(v, k)\|_k = \|v\|,$$

where the norm on the right is in $\mathscr{B}$. We have then made $\mathscr{B}_k$ into a B-space.

We say that two pairs $(v, k) \in \mathscr{B}_k$ and $(w, p) \in \mathscr{B}_p$, $p \geq k$, are *equivalent* if $\mathbf{T}^{k-p} v = w$ in $\mathscr{B}$ (for $p = k$ we set $\mathbf{T}^0 = 1$). We now define $\mathfrak{B}$ as the union of the spaces $\mathscr{B}_k$, $k = 0, \pm 1, \ldots$, taking the elements of $\mathfrak{B}$ to be equivalence classes of pairs. Observing that any two pairs can be represented as elements of the same space $\mathscr{B}_k$:

$$(w, k) = (T^{-1} w, k+1) = (T^{-2} w, k+2) = \ldots,$$

we can define, in a natural way, a linear operation in $\mathscr{B}$ that converts it into a linear space over $\mathbb{C}$. We identify the original space $\mathscr{B}$ with $\mathscr{B}_0$. We can extend T to the whole space $\mathfrak{B}$ by setting

$$T(v, k) = (v, k+1). \tag{1}$$

Since when v and $T v = w$ belong to $\mathscr{B}$, we have

$$T(v, 0) = (v, 1) = (T^{-1} v, 1) = (w, 0),$$

i.e. $T(v, 0) = (T v, 0)$, the definition (1) really provides an extension of T and is consistent with the definition $T^{-1}(v, k) = (T^{-1} v, k)$.

Remark 1. The preceding argument is equivalent to the following proposition: an element $v \in \mathscr{B} \equiv \mathscr{B}_0$ belongs to the domain of the operator T: $\mathscr{B} \to \mathscr{B}$ if and only if there is a pair $(w, -1)$ that is equivalent to $(v, 0)$. In fact, if there is such a pair then $T(v, 0) = T(w, -1) = (w, 0)$, i.e. $T v = w$. But if $v \in \mathfrak{D}(T)$, $T v = w$, $v = T^{-1} w$, then $(v, 0) = (T^{-1} w, 0) = (w, -1)$.

We now introduce an additional structure on $\mathfrak{B}$ that substitutes for the topologization and is a variation of countable normability.

Definition. *A set* $\mathfrak{M} \subset \mathfrak{B}$ *is said to be* bounded *if there is a* k *such that* $\mathfrak{M}$ is bounded in $\mathfrak{B}_k$.

The linear space $\mathfrak{B}$ with the supplementary structure will be called an *O-space*[3] (space with bounded sets).

Remark 2. A better known procedure is to introduce the inductive limit topology in $\mathfrak{B}$ (see [10], [43]) by the chain of spaces

$$\ldots \subset \mathscr{B}_{-2} \subset \mathscr{B}_{-1} \subset \mathscr{B}_0 \subset \mathscr{B}_1 \subset \mathscr{B}_2 \subset \ldots,$$

but this introduces additional complications when T^{-1} is not completely continuous.

Remark 3. If T is an operator generated by a differential operation, then $\mathscr{B}_k$, $k \geq 1$, is a space of generalized functions, and $\mathscr{B}_{-k}$ is a space of smooth functions. This system of notation conflicts with the traditional terminology in which negative indices enumerate spaces of generalized functions (see [2]), but is more natural in our approach.

We now take up the construction of an algebra of operators on $\mathfrak{B}$ which serves as a framework for the construction of the operational calculus. The class of linear transformations of $\mathfrak{B}$ on itself which carry each set that is

---

[3] *O* is the first letter of the Russian word for "bounded" (Translator).

bounded in $\mathscr{B}_p$ to a set bounded in $\mathscr{B}_{p+k}$ (possibly $k<0$) will be denoted by $\mathfrak{A}_k$. Each set $\mathfrak{A}_k$ has the structure of a B-space: linear operations are defined in the natural way and for $\mathbf{A}\in\mathfrak{A}_k$

$$\|\mathbf{A}\|_k = \sup_p \sup_{u\in\mathscr{B}_p} \frac{\|\mathbf{A}u\|_{p+k}}{\|u\|_p},$$

where the right-hand side is finite by hypothesis.

The union of the $\mathfrak{A}_k$ forms an algebra $\mathfrak{A}$, whose additive group has the structure of an O-space. In fact, because of the existence of the natural embedding $\mathfrak{A}_k \subset \mathfrak{A}_p$, $p\geq k$, when $\mathbf{A}\in\mathfrak{A}_k$ and $B\in\mathfrak{A}_p$ the linear combination

$$\alpha\mathbf{A}+\beta\mathbf{B}\in\mathfrak{A}_p; \qquad \alpha,\beta\in\mathbb{C},$$

is defined, and a set $\mathfrak{P}\subset\mathfrak{A}$ is bounded if it is bounded in some $\mathfrak{A}_k$.

As the next step toward realizing the plan presented in 2.2, Chapter I, whose realization lets us define, for $\mathbf{A}\in\mathfrak{A}$, functions $\varphi(\mathbf{A})$ with values in the algebra $\mathfrak{A}$, we need to consider functions of a complex variable with values in $\mathfrak{A}$.

Remark 4. In applications of the operational calculus with whose construction we are concerned, one usually considers the case $\mathbf{A}=\mathbf{T}$, i.e. a given element of $\mathfrak{A}$, for which the calculus is constructed, coincides with the operator that is used to construct $\mathfrak{B}$. However, in principle this is not necessarily the case.

We first consider the resolvent of an element $\mathbf{A}\in\mathfrak{A}$. We introduce, in the usual way, a function of the complex parameter $\lambda\in\mathbb{C}$:

$$\mathbf{R}_\lambda(\mathbf{A})=(\lambda\mathbf{E}-\mathbf{A})^{-1}\equiv(\lambda-\mathbf{A})^{-1}, \tag{2}$$

defined for the values of λ for which an element of $\mathfrak{A}$ is defined by (2).

Definition. *A set $\sigma_k(\mathbf{A})\subset\mathbb{C}$, with the property that for every $\lambda\in\mathbb{C}\setminus\sigma_k(\mathbf{A})$ the function $\mathbf{R}_\lambda(\mathbf{A})$ exists, belongs to $\mathfrak{A}_k$, and the inequality*

$$\|\mathbf{R}_\lambda(\mathbf{A})\|_k \leq M(\mathbf{A},k)=\text{const} \tag{3}$$

(the norm on the left is taken in $\mathfrak{A}_k$) is satisfied uniformly in λ, is called k-spectral for A.

Notice that σ_k is not assumed to be closed.

Proposition 1. *If $\lambda\in C\setminus\sigma_k(\mathbf{A})$, $\mathbf{A}\in\mathfrak{A}_l$, then the function $\lambda(\lambda-\mathbf{A})^{-1}$ is bounded in $\mathfrak{A}_{k+l}$.*

For the proof, it is enough to observe that

$$\lambda(\lambda-\mathbf{A})^{-1}=1+\mathbf{A}(\lambda-\mathbf{A})^{-1}.$$

By analogy with the usual definition of holomorphy for functions with values in a B-space, a function $\varphi(z)$, $z \in \mathbb{C}$, with values in an O-space $\mathfrak{A}$, is said to be *holomorphic* in an open set $\Omega \subset \mathbb{C}$ if it is holomorphic in Ω as a function with values in the B-space $\mathfrak{A}_k$ for some k.

Proposition 1 lets us establish the standard property of the resolvent:

Proposition 2. *At interior points of the set* $\mathbb{C} \backslash \sigma_k(\mathbf{A})$ *the function* $R_\lambda(\mathbf{A})$ *is a holomorphic function of* λ *with values in* $\mathfrak{A}$.

Proof. When λ_0 belongs to the interior of $C \backslash \sigma_k(\mathbf{A})$, and λ is sufficiently close to λ_0, we may use the identity

$$\frac{1}{\lambda_0 - A} - \frac{1}{\lambda - A} = (\lambda - \lambda_0) \frac{1}{\lambda_0 - A} \frac{1}{\lambda - A}$$

and the boundedness of multiplication in $\mathfrak{A}$ to obtain the existence of $(d/d\lambda) R_\lambda(\mathbf{A})$ in an appropriate space $\mathfrak{A}_l$. $\square$

We now turn to the description of a rather large class $\mathfrak{F} \equiv \mathfrak{F}(\mathbf{A})$ of complex functions that have the property that $\varphi(\mathbf{A}) \in \mathfrak{A}$ is defined for $\varphi(z) \in \mathfrak{F}$. As we should expect, the set $\mathfrak{F}$ is also an algebra.

Let $\mathbf{A} \in \mathfrak{A}$ and let the set $\sigma_q \equiv \sigma_q(\mathbf{A}) \subset \mathbb{C}$ be q-spectral for $\mathbf{A}$ for some value q. Let G be an open set in $\mathbb{C}$ and $\bar{\sigma}_q \subset G_\varepsilon \subset G$, where $\bar{\sigma}_q$ is the closure of σ_q and G_ε is the set of interior points of G whose distance from the boundary of G is greater than $\varepsilon > 0$. We may suppose without loss of generality that zero is not in G and that consequently z^{-1} is holomorphic in G. A domain G with these properties with respect to $\mathbf{A}$ is called $\mathbf{A}$-*admissible*.

For each integer k, we denote by $\mathfrak{F}_k$ the linear manifold of complex functions that are holomorphic at interior points of G, continuous in $\bar{G}$, and such that for $\varphi \in \mathfrak{F}_k$

$$|\varphi(z)| \leq M_\varphi |z|^k \quad \text{for } z \in \bar{G}. \tag{3}$$

If we now define the norm for $\varphi \in \mathfrak{F}_k$ by

$$|\varphi, \mathfrak{F}_k| = \sup_{z \in \bar{G}} |z^{-k} \varphi(z)|,$$

then $\mathfrak{F}_k$ becomes a B-space. The union of all the $\mathfrak{F}_k$, $k = 0, \pm 1, \pm 2, \ldots$ becomes an algebra in the natural way, with the additional structure of an O-space. We shall call it the $\mathbf{A}$-*admissible algebra* $\mathfrak{F} \equiv \mathfrak{F}(G)$.

Each element $\varphi(z) \in \mathfrak{F}$ admits a canonical representation as a contour integral. Let us now describe this. Let γ be a curve (generally speaking, unbounded), lying in $G \backslash G_\varepsilon$ and bounding a domain G_γ such that

$$G_\varepsilon \subset G_\gamma \subset G$$

If K_r is the disk $|z| \leq r$, then γ_r is the part of γ that lies in K_r. Suppose that for each $r > 0$ the curve γ_r is rectifiable and that there is a constant $c > 0$ such that

$$\text{length of } \gamma_r \leq cr.$$

If $\gamma_r \cup \gamma_r'$ is the boundary of the domain $G_{\gamma,r} = G_\gamma \cap K_r$ consisting of γ_r and the corresponding parts γ_r' of the circle $|z| = r$, the representation

$$\varphi(z) = \frac{z^p}{2\pi i} \int_{\gamma_r \cup \gamma_r'} \frac{\varphi(\zeta)}{\zeta^p(\zeta - z)} d\zeta \tag{4}$$

is valid for every $z \in G_\varepsilon \cap K_r$, and every function $\varphi \in \mathfrak{F}$, with corresponding orientations of the curves, for every integer p.

When $\varphi \in \mathfrak{F}_k$ and $p \geq k + 2$, we may take the limit on the right-hand side of (4) as $r \to \infty$, obtaining

$$\varphi(z) = \frac{z^p}{2\pi i} \int_\gamma \frac{\varphi(\zeta)}{\zeta^p(\zeta - z)} d\zeta, \tag{5}$$

since the integral over γ_r' tends to zero because of the rapid decrease of the integrand.

According to Proposition 2, the resolvent $\mathbf{R}_\lambda(\mathbf{A})$ is a holomorphic function of λ with values in A for a suitably defined domain $G \subset \mathbb{C}$. Here the holomorphy of $\mathbf{R}_\lambda$ means holomorphy in a B-space and we may operate accordingly with $\mathbf{R}_\lambda$, in particular considering integrals involving $\mathbf{R}_\lambda$ along curves that lie in G. Therefore the formula

$$\varphi(\mathbf{A}) = \frac{\mathbf{A}^p}{2\pi i} \int_\gamma \frac{\varphi(\zeta)}{\zeta^p(\zeta - \mathbf{A})} ds, \tag{6}$$

where the integral has the same meaning as in (5), and automatically exists for sufficiently large p, defines an element of $\mathfrak{A}$.

Theorem. *Under the preceding hypotheses, formula* (6) *defines a bounded linear homomorphism* $h_\mathbf{A}$:

$$h_\mathbf{A}: \mathfrak{F} \to \mathfrak{A}; \qquad \varphi(z) \to \varphi(\mathbf{A})$$

of the **A**-*admissible algebra* $\mathfrak{F}$ *of complex functions into the algebra* $\mathfrak{A}$; *here*

$$h_\mathbf{A}(z) = \mathbf{A}; \qquad h_\mathbf{A}(1) = 1.$$

Proof. The boundedness and linearity of the homomorphism given by (6) are rather obvious. Let us verify that $\varphi(\mathbf{A})$ in (6) is independent of p for sufficiently large values of p.

In the notation of formula (4) we can write, for some $r > 0$, the chain of equations

$$A^p \int_{\gamma_r} \frac{\varphi(\zeta)}{\zeta^p(\zeta - A)} d\zeta - A^{p+1} \int_{\gamma_r} \frac{\varphi(\zeta)}{\zeta^{p+1}(\zeta - A)} d\zeta$$

$$= A^p \int_{\gamma_r} \frac{(\zeta - A)\varphi(\zeta)}{\zeta^{p+1}(\zeta - A)} d\zeta = A^p \left(\int_{\gamma_r \cup \gamma_r'} - \int_{\gamma_r'} \right) \frac{\varphi(\zeta)}{\zeta^{p+1}} d\zeta = -A^p \int_{\gamma_r'} \frac{\varphi(\zeta)}{\zeta^{p+1}} d\zeta.$$

At the last step we used the fact that the first integral in the parentheses is zero because $\varphi(\zeta)\zeta^{-p-1}$ is holomorphic in the corresponding domain. If we now take limits as $r \to \infty$ and observe that the integral over γ_r' tends to zero, we obtain

$$A^p \int_{\gamma} \frac{\varphi(\zeta)}{\zeta^p(\zeta - A)} d\zeta = A^{p+1} \int_{\gamma} \frac{\varphi(\zeta)}{\zeta^{p+1}(\zeta - A)} d\zeta. \tag{7}$$

It follows from (7) that $h_A(z\varphi) = A h_A(\varphi)$, $A^{-1} = h_A(z^{-1})$. The verification of the equation

$$h_A(\varphi_1 \varphi_2) = h_A(\varphi_1) h_A(\varphi_2)$$

is carried out in the usual way by using the representations of φ_1 and φ_2 by integrals of the form (6) over appropriate curves γ_1 and γ_2, and Hilbert's identity (see [5, pp. 569–570]). $\quad\square$

Remark 5. In studying the convergence of the integrals that arise in the discussion it is sometimes useful to use the inversion $\zeta \to 1/\zeta$, supposing (as is always admissible) that γ does not pass through zero. Then the image of γ is a bounded curve.

Remark 6. It is rather obviously possible to extend the preceding discussion to a set of commuting operators $A_1, \ldots, A_m$ (see [42]). The more complicated problem of defining an operator function of the form $\varphi(t, A)$, where t is an additional parameter, will be discussed in the next section in terms of a specific example.

§2. Some Examples

Let us discuss the possibility of applying our operational calculus to some interesting operator equations. As usual, we concentrate on the simplest examples, which clarify the theoretical side of the question.

For the equation

$$Lu \equiv (D_t - A)u = f, \quad t \in [0, b], \tag{1}$$

we use the formula that, when A is a number, gives the solution of a Cauchy problem:

$$u(t) = \int_0^t e^{(t-\tau)A} f(\tau) d\tau. \tag{2}$$

Considering A as a parameter, and desiring to give a meaning to (2) in the framework of an operational calculus containing functions of $\mathbf{A}$, we must, first of all, consider the function $\varphi(z) = e^{\eta z}$, where $\eta \in [0, b]$. However, for such a φ an inequality of the form (3), §1, can be used (for arbitrary k) only in a half plane $\operatorname{Re} z \leq M$, where M is any number. Consequently if we wish φ to belong to the A-admissible algebra $\mathfrak{F}$, a set that is k-spectral for A for some k should be a half plane $\operatorname{Re} z \geq N$, where N is again an arbitrary number. If this condition is satisfied the function $\varphi(\mathbf{A}) = e^{(t-\tau)\mathbf{A}}$ defines an element of the algebra $\mathfrak{A}$, and for every function $f \in \mathfrak{B}$ that is integrable with respect to τ the formula (2) defines an element $u \in \mathfrak{B}$ which is a solution of (1).

In fact, integration with respect to τ (or differentiation with respect to t to verify that $u(t)$ is a solution of (1)) can be carried out in the usual way, as for functions of a numerical parameter with values in a B-space. If the operator $\mathbf{A}$ (or, more precisely, its resolvent) has the property that if f belongs to $\mathscr{B}_k$ then $e^{\eta \mathbf{A}} f \in \mathscr{B}_{k+1}$, then the element $u(t)$ defined by (2) will belong to $\mathscr{B}_{k+1}$.

A solution of (1), obtained by using the operational calculus, will be called an *O-solution*. Since our construction ensures that the inequality

$$\|u\| \leq c \|f\| \tag{3}$$

is satisfied (the norm on the left is the norm in $\mathrm{I\!H}_t \otimes \mathscr{B}_{k+1}$; that on the right, in $\mathrm{I\!H}_t \otimes \mathscr{B}_k$), the solution is unique in the same sense.

Remark. If in our hypotheses the space $\mathscr{B}_0$ is our standard space $\mathrm{I\!H}_x$, and if there corresponds to a sufficiently smooth right-hand side f a classical (sufficiently smooth) solution of (1), then the O-solution will be a strong solution in the sense of the definition in Chapter II. In fact, it is always possible to construct smooth approximations $\{f_i\}$ to the right-hand side, and then, under our hypotheses, because of (3) the convergence of f_i to f will correspond to the convergence of the sequence of smooth solutions u_i to u.

If we turn now to specific differential operators that are suitable for substitution into (2) as the operator $\mathbf{A}$, and if we also require that these are not M-operators, i.e. that it is really necessary to use the framework of Chapters V and VI, we see at once how extraordinarily stringent the conditions imposed on $\mathbf{A}$ really are. Thus if $x \in [0, b]$ then the operator aD_x^2, considered in $\mathrm{I\!H}_x$ with the Cauchy conditions $u|_{x=0} = u'_x|_{x=0} = 0$ is unavailable for any $a \neq 0$ (real or complex). The operator aD_x, determined by the condition $u|_{x=0} = 0$, is available only for real $a > 0$. It is true that when the preceding condition is satisfied we can replace aD_x by any operator of the form

$$\mathbf{A} \equiv aD_x + \mathbf{B},$$

where $\mathbf{B}$ is an M-operator the location of whose spectrum in the complex plane is subject to appropriate restrictions.

We discuss this operator in detail, considering only the case $\mathbf{B}=0$. The use of operational calculus for $\mathbf{B} \neq 0$ in combination with the constructions of Chapter V is suggested as an exercise (one may, for example, take $\mathbf{B}$ to be the Π-operator generated by the operation $-D_1^2-D_2^2$).

Thus if $\mathbf{A}$ is generated in the space $\mathscr{B} \equiv \mathscr{B}_0 = \mathrm{IH}_x$ by the operation aD_x, $a \neq 0$, and the condition $u|_{x=0}=0$, then

$$\mathbf{R}_\lambda(\mathbf{A})\,g \equiv (\mathbf{A}-\lambda)^{-1}\,g = a^{-1}\int_0^x e^{a^{-1}(x-\xi)\lambda}\,g(\xi)\,d\xi. \tag{4}$$

From this and the requirement (above) on the spectral set of $\mathbf{A}$, it follows immediately that $a>0$. It also follows from (4) that in the case under consideration $(\mathbf{A}-\lambda)^{-1}$ acts boundedly from $\mathscr{B}_0$ to $\mathscr{B}_{-1}$. Then the operator $e^{\eta\mathbf{A}}$ defined, by (5), § 1, by the formula

$$e^{\eta\mathbf{A}} = \frac{\mathbf{A}}{2\pi i}\int_\gamma \frac{e^{\eta z}}{z(z-\mathbf{A})}\,dz,$$

where γ is a vertical line, acts boundedly from $\mathscr{B}_0$ to $\mathscr{B}_0$ (or, equivalently, from IH_x to IH_x) and formula (2) provides a solution of (1). In this case the norms in (3) are norms in IH, and, as is easily seen (see the remark on inequality (3)), the O-solution is a strong solution, and the corresponding operator $\mathbf{L}: \mathrm{IH} \to \mathrm{IH}$ is proper.

The modification of the preceding construction for the inverse Cauchy problem in t $(u|_{t=b}=0)$ is evident: the half planes $\mathrm{Re}\,z \leq c$ and $\mathrm{Re}\,z \geq c$ exchange roles.

We turn now to the case when the operator $D_t: \mathrm{IH} \to \mathrm{IH}$ in (1) is given by the conditions

$$\mu u|_{t=0} - u|_{t=b} = 0. \tag{Γ_μ}$$

The solution u of the corresponding equation (for a numerical parameter $\mathbf{A}$) is now given by formula (11), § 2, Chapter III (with λ replaced by $\mathbf{A}$). If we do not wish to subject $\mathbf{A}$ to restrictions similar to those used in the Cauchy problem, then a separate discussion of the exponential $e^{\eta\mathbf{A}}$ is not helpful (see the inequalities used in the proof of Lemma 2, § 1, Chapter V) and we need a modification of the construction of § 1.

Let $\varphi(z,t\,|\,f)$, for some given element $f \in \mathscr{B}$, be a function of z, holomorphic for $z \in Q \subset \mathbb{C}$, with values in $\mathscr{B}$ and depending on $t \in [0,b]$ as a parameter. If, in addition, the inequality

$$\|\varphi\| \leq c_\varphi |z|^p \tag{5}$$

is satisfied for $z \in Q$, where the norm on the left is in $\mathscr{B}$ (i.e., in an appropriate B-space $\mathscr{B}_k$), then φ, as a function of z, has a canonical representation analogous to that in § 1:

$$\varphi(z,t|f) = \frac{z^k}{2\pi i} \int_\gamma \frac{\varphi(\zeta,t|f)}{\zeta^k(\zeta - z)} \, d\zeta \in \mathfrak{B}, \tag{6}$$

where $k \geq p+2$ and γ (in general, unbounded) satisfies corresponding requirements. Since the operator $(\zeta - A)^{-1} = R_\zeta(A)$ is defined for $\zeta \in \gamma$ under corresponding assumptions about the spectral set of A (analogous to those used in § 1), formula (6) lets us define an element of $\mathfrak{B}$ of the form

$$\varphi(A,t|f) = \frac{A^k}{2\pi i} \int_\gamma \frac{\varphi(\zeta,t|f)}{\zeta^k(\zeta - A)} \, d\zeta. \tag{6'}$$

We would like to apply the preceding discussion to the case when

$$\varphi(z,t|f) = (\mu - e^{bz})^{-1} \left(\mu \int_0^t e^{(t-\tau)z} f(\tau) \, d\tau + e^{bz} \int_t^b e^{(t-\tau)z} f(\tau) \, d\tau \right). \tag{7}$$

Under the assumption that $|\mu - e^{bz}| \geq \delta > 0$, the discussion in § 1, Chapter V, immediately yields an inequality of the form (5). If M_σ is the set of zeros of the function $\mu - e^{bz}$:

$$M_\sigma = b^{-1} \{\ln|\mu| + i(\arg\mu + 2k\pi)\}, \quad k = 0, \pm 1, \pm 2, \dots,$$

then the requirements on γ that are used in (6) and (6') are evident, and our results can be stated as a theorem:

Theorem. *If, for given μ and $\delta > 0$, the set of points $\{z : d(z, M_\sigma) \leq \delta\}$ belongs to the k-resolvent set of A, then (6'), with $\varphi(z,t|f)$ defined by (7), defines an O-solution of (1) under conditions (Γ_μ).* $\square$

A further discussion along the lines of that used for the Cauchy problem shows that under corresponding assumptions about the resolvent of A the O-solution is a strong solution.

Our discussion of formulas (2) and (7) from the operational point of view again emphasizes the profound difference between the Cauchy problem and the nonlocal problem (Γ_μ). The Cauchy problem, when it belongs to a proper operator, is extremely stable with respect to a certain class of perturbations, but the (Γ_μ) problem lets us use a significantly larger class of operators A.

We have confined ourselves to the examination of the simplest examples. How to apply the same methods for the equations and problems considered in Chapter VI is fairly clear, but the difficulty of analyzing specific problems naturally increases.

§ 3. The Necessity for Restrictions on the Resolvent

A characteristic feature of the discussion in Chapters V and VI, where the operator equations involved Π-operators (or M-operators) A_k, was the definitive nature of the results in the sense of the necessity and sufficiency of the requirements (at least in the majority of cases) on the resolvents of the A_k (on the structure of their point spectra) that guaranteed the proper character of the operator described by various boundary conditions on t.

Evidently the application of the operational calculus permits the construction of an O-solution of the operator equation under various sufficiency conditions on the operator A (or operators A_k). However, the construction employed does not immediately yield a method for determining the necessity of the conditions imposed on A for existence or uniqueness, for example, of the strong solution of the problem. Even for the simplest equation (1), § 2, when we have at our disposal such a classical result as the Hille-Yosida theorem (see [21]) (which gives necessary and sufficient conditions on the resolvent of A that guarantee the existence of an operator S_t, bounded in the requisite sense, that provides the solution $u(t) = S_t u_0$ of the problem $Lu = 0$, $u|_{t=0} = u_0$), an attempt to apply the result directly to our situation leads to difficulties.

Not having any general method for studying this question, we limit ourselves to the discussion of some simple examples. Let

$$L(D) \equiv D_t + a D_x,$$

where a is a complex number and, as usual, $t \in [0, b]$, $x \in [0, 2\pi]$. The operator D_x: $\mathbb{H}_x \to \mathbb{H}_x$ is now determined by means of the Cauchy condition

$$u|_{x=0} = 0. \tag{1}$$

If D_t: $\mathbb{H}_t \to \mathbb{H}_t$ is defined by starting from the condition

$$\mu u|_{t=0} - u|_{t=b} = 0,$$

then when $\mu \neq 0$, ∞ the fact that the associated operator L: $\mathbb{H} \to \mathbb{H}$ is proper (in the wide sense) follows immediately, for every a, from the discussion in Chapter V, if the roles of x and t are interchanged. Hence only the case $\mu = 0$, ∞ is now of interest. Consider, for definiteness, the condition

$$u|_{t=0} = 0. \tag{2}$$

Then it follows from the results of § 2 that the operator L is proper for $a > 0$ (this condition ensures that the resolvent of $A \equiv -aD_x$ has the necessary properties). We establish the necessity of this requirement by showing that its failure leads to the unboundedness of the operator L^{-1}: $\mathbb{H} \to \mathbb{H}$.

In the same way as it followed from the discussion in §2 that when $a>0$ we had the typical situation for Cauchy problems, $\rho L = C$, it will follow from the construction given below that when the preceding condition is violated the continuous spectrum $C\sigma L$ fills the complex plane $\mathbb{C}$.

Thus, in the equation

$$\mathbf{L}(D)u = f,$$

considered under conditions (1) and (2), we choose a sequence of right-hand sides f_v of the form

$$f_v(t,x) = e^{vx} - e^{-avt}. \tag{3}$$

The corresponding family of solutions u_v have the form

$$u_v = \frac{1 - e^{-avt}}{av}(e^{vx} - 1). \tag{4}$$

The relation of interest,

$$\|u_v\|/\|\mathbf{L}u_v\| = \|u_v\|/\|f_v\| \to \infty,$$

for some sequence of values $\{v_k\}$, $k \to \infty$, will automatically be satisfied if, for some given values x and $t(x>0,\ t>0)$ the condition

$$\eta_k = \frac{|e^{v_k x} e^{-av_k t}|}{|e^{v_k x}| + |e^{-av_k t}|} \to \infty \tag{5}$$

is satisfied as $k \to \infty$. Let $a = a' + ia''$, $v_k = v_k' + iv_k''$. Then

$$\eta_k = \frac{e^{v_k' x} e^{(-a'v_k' + a''v_k'')t}}{e^{v_k' x} + e^{(-a'v_k' + a''v_k'')t}}.$$

Since, for the relation (5) that we need, it is necessary that both exponentials tend to $+\infty$ as $k \to \infty$, we may set $v_k' = k$. Then, as expected, (5) cannot be satisfied if $a' > 0$ and $a'' = 0$. If $a' < 0$ and $a'' = 0$, then we can obtain (5) if $v_k'' = 0$. However, if $a'' \neq 0$ then, for example, $v_k'' = (a'')^{-1}k(1+a')$ (which implies $-a'k + a''v_k'' = k$) again yields (5). This completes our discussion.

It is natural to ask what motivates the choice of the family (3), or equivalently of (4). A rather vague answer is given by the fact that using the family of functions $e^{vx} - 1$ is very convenient for explaining the nature of an operator D_x generated by Cauchy conditions.

We can illustrate this remark by considering a somewhat complicated variant of the operation $\mathbf{L}(D)$:

$$\mathbf{L}(D) = D_t + aD_x^2,$$

where D_t is again defined by (2), and D_x^2 by the condition

$$u|_{x=0}=u'_x|_{x=0}=0.$$

Then, taking u_v in the form

$$u_v=(av)^{-1}(1-e^{-av^2t})(e^{vx}-1-vx)$$

and correspondingly

$$f_v=\mathbf{L}(D)u_v=ve^{vx}-ve^{-av^2t}(1+vx),$$

we obtain, instead of (5), the following expression for η_k:

$$\eta_k=\frac{e^{-av_k^2t}e^{v_kx}}{e^{-av_k^2t}+e^{v_kx}}.$$

Retaining the previous notation, setting $v'_k=k$, we can establish the unboundedness of the operator $\mathbf{L}^{-1}:\,\mathbb{H}\to\mathbb{H}$ by considering the cases

1) $a'<0;\ v''_k=0,$
2) $a'=0,\ a''\neq0;\ v''_k=-a''k,$
3) $a'>0,\ a''>0;\ v''_k=-2k,$
   $a'>0,\ a''<0;\ v''_k=2k.$

With this we conclude the discussion of the question of the necessity of the restrictions imposed on the resolvent in §2.

Concluding Remarks

The author hopes that an attentive reader will have obtained from the preceding chapters an idea of the nature of the general questions that arise in the study of boundary value problems by the methods of the theory of linear operators, and will be able to use these methods for the study of the special class of representative entities that make it possible to give sufficiently meaningful answers to some of these questions.

The illustrative situations that we have considered allow a deeper understanding of many special phenomena connected with corresponding parts of mathematical analysis. As we observed at the end of Chapter VI, a similar approach can be used to study linear problems with a small (or large) parameter, as well as for equations that degenerate on the boundary or are of mixed type. It is assumed that readers can, if necessary, adapt these methods to their own problems by using the ideas presented in Chapters V and VI.

The author believes that constructions, as used in this monograph, supplement, in an essential way, the point of view on linear differential operators that is presented in [31].

The fundamental question that arises in connection with our presentation is the question of the extent to which the properties of models are preserved in more general situations. Here it would be desirable to have a reasonable theory of perturbations which would let us pass from a parallelepiped (torus) to a domain obtained by some small deformation, or from a Π-operator to a neighboring operator with variable coefficients.

One should not, of course, think that such a theory of perturbations would make it possible to start from our models and obtain results that would immediately apply to the general linear differential operation with variable coefficients, considered in an arbitrary domain. However, it should provide useful heuristic ideas, both about the formulation of problems and the nature of the expected results.

To make the corresponding range of problems specific, to a certain extent, we may suggest, for example, the following questions.

For what classes of operations with variable coefficients are there boundary conditions such that the proper operators generated by them are M-operators? C-operators? How much and in what way does the possibility of choosing the required boundary conditions depend on the structure of the domain in which the problem is discussed?

Up to now, however, it is not clear how to show, even for the Laplace operator in a disk (or in a rectangle) that it does not generate a proper operator that is a C-operator (or how to prove the contrary).

It follows from the results of [31] that, for example, for an ultrahyperbolic operation (see [14]) there must exist a proper operator L generated by it, such that $L^{-1}\colon \mathbb{H} \to \mathbb{H}$ is not only bounded, but also completely continuous. The question of how to describe such an operator in terms of boundary conditions remains open. The construction used for this purpose in 1.4, Chapter VI, in combination with the choice of μ_s to guarantee the growth of the constants in inequalities of the form (11), §1, Chapter VI, encounters a difficulty connected with the necessity of a supplementary investigation of the properties of the resulting basis.

The question formulated above for an ultrahyperbolic operator is evidently a special case of the general problem of the dependence (for a given differential operation $L(D)$) of the properties of L^{-1} on the choice of the boundary conditions for L.

As another version of this problem we may propose the investigation of how the smoothness of the solution $u = L^{-1}f$ depends on the nature of the (uniform) boundary conditions, for $f \in \mathbb{H}$. A useful example in this connection is discussed in [36].

There is a certain interest in applying the methods we have presented to systems of equations. Here there immediately arises the question of considering operator functions of several complex variables. In this direction there are still, however, only very isolated results.

There is also the obvious possibility of going over (within the framework of our ideas) to the noncompact case as is done in the classical case by the Fourier-Laplace transform, as well as by applying the operational calculus of Chapter VIII.

In completing this exposition of "general questions", the author wishes to express the hope that the further study of them, the appropriateness and utility of which he has attempted to demonstrate, will eventually lead to a "general theory" of linear partial differential boundary value problems.

Appendix 1
On Some Systems of Equations Containing
a Small Parameter

As this book was being prepared for the press, one step was taken in the plan for later investigation of the proposed scheme for the analysis of simplified versions of some problems that arise in the linear theory of boundary value problems. Here it is a question of systems of equations that arise for a special linearization of the two-dimensional Navier-Stokes equations. We present an exposition of the corresponding construction, following [44].

§1. Formulation of the Problem

Let $U=(u,v)$ be a two-dimensional vector function of the variables x, y; $p=p(x,y)$ a scalar; and

$$\varepsilon \Delta U + \mathbf{K} U + \operatorname{grad} p = 0, \tag{1}$$

$$\operatorname{div} U = 0 \tag{2}$$

the classical system of equations whose solutions describe the stationary plane flow of a viscous incompressible fluid [2]. The convective term in (1) can be written in either of the forms

$$\mathbf{K} U = \begin{pmatrix} u\dfrac{\partial u}{\partial x} + v\dfrac{\partial u}{\partial y} \\ u\dfrac{\partial v}{\partial x} + v\dfrac{\partial v}{\partial y} \end{pmatrix}, \tag{3}$$

or

$$\mathbf{K} U = \begin{pmatrix} -v\left(\dfrac{\partial v}{\partial x} - \dfrac{\partial u}{\partial y}\right) \\ u\left(\dfrac{\partial v}{\partial x} - \dfrac{\partial u}{\partial y}\right) \end{pmatrix}. \tag{4}$$

In the latter case we must replace p by the scalar $\gamma = p + \frac{1}{2}(u^2 + v^2)$.

To linearize **K** as a group of minor terms in (1), we may use either a hyperbolic or an elliptic operator of the first order. For example, setting $v = 0$, $u = 1$ in (4) (having in view the coefficients of the first derivatives), and adding (2) to the first row of (1), we obtain, as **K**, the classical Cauchy-Riemann operator. To obtain a hyperbolic (split) operator it is enough to set $u = v = 1$ in (3).

We shall be interested in the behavior, as $\varepsilon \to 0$, of the solution of the linearized system (1)–(2), considered as a system with constant coefficients in a rectangle, with a special choice of boundary conditions. Here we concentrate on the case when **K** is the Cauchy-Riemann operator. The corresponding results are more evident in the hyperbolic case.

We start by considering the truncated system obtained from (1) for $p \equiv 0$ (then equation (2) drops out). This makes it possible to emphasize the characteristics of the complete system (1)–(2).

We now introduce a method for applying the familiar construction described in Chapters V and VI. We consider the operator equation

$$\mathbf{L}_\varepsilon U_\varepsilon \equiv \{\varepsilon(D_x^2 + \mathbf{M}) + \mathbf{B} D_x - \mathbf{A}\} U_\varepsilon = 0, \tag{L_ε}$$

where $x \in [0, b]$, D_x is the operation of differentiation with respect to x, and **A**, **B**, and **M** are Π-operators that commute with D_x (the role of t in the cited chapters is now played by x).

Together with the equation (L_ε) we consider boundary conditions for x (for $x = 0, b$), which we write symbolically in the form

$$\Gamma U_\varepsilon = v, \tag{Γ}$$

and which we suppose are meaningful for every $U \in \mathfrak{D}(\mathbf{L}_\varepsilon)$. Besides (L_ε) we consider the unperturbed equation

$$\mathbf{L}_0 U \equiv \{\mathbf{B} D_x - \mathbf{A}\} U = 0, \tag{L_0}$$

with the associated boundary conditions

$$\Gamma_0 U = v_0 \tag{Γ_0}$$

(conditions (Γ_0) are obtained as usual by dropping part of conditions (Γ)).

Fundamental hypothesis. Problems $(L_\varepsilon)–(\Gamma)$ and $(L_0)–(\Gamma_0)$ have unique solutions for all v and v_0 and the solutions satisfy the inequalities

$$\|U_\varepsilon\| \leq C \|v\|', \quad \|U\| \leq C \|v_0\|', \tag{5}$$

(with the use of appropriate norms). The constants C in (5) are independent of the choice of solution, and the first inequality is satisfied uniformly in ε.

This proposition is essential for establishing the connections that are of interest here between the original and perturbed problems. Observe also that the decomposition of problems $(L_\varepsilon)-(\Gamma)$ and $(L_0)-(\Gamma_0)$ into infinite chains of ordinary differential equations can be effected in the usual way. We have the following theorem.

Theorem 1. *A necessary and sufficient condition for the validity of the fundamental hypothesis is that each of the ordinary equations (L_ε, s) and $(L_{0,s})$, mentioned above, is uniquely solvable for all v_s and all $v_{0,s}$, and that the inequalities*

$$\|U_{\varepsilon,s}\| \le C|v_s|, \qquad \|U_s\| \le C|v_{0,s}| \tag{6}$$

are satisfied uniformly in ε and s, or in s, respectively.

This proved in the usual way.

Definition. When the fundamental hypothesis is satisfied, the $(L_\varepsilon)-(\Gamma)$ *problem degenerates regularly into the $(L_0)-(\Gamma)$ problem if*

$$\|U_\varepsilon - U\| \to 0 \quad as \quad \varepsilon \to 0.$$

Theorem 2. *When the fundamental hypothesis is satisfied, the $(L_\varepsilon)-(\Gamma)$ problem degenerates regularly into the $(L_0)-(\Gamma_0)$ problem as $\varepsilon \to 0$, if and only if, for each given $s \in S$,*

$$\|U_{\varepsilon,s} - U_s\| \to 0 \quad as \quad \varepsilon \to 0. \tag{7}$$

Proof. Let us verify the sufficiency of the conditions. Take an arbitrary $\delta > 0$. By the fundamental hypothesis there is an N such that we will have, in the representations for U_ε and U,

$$\|\sum_{|s|>N} U_s \varphi_s\| < \frac{\delta}{3}, \qquad \|\sum_{|s|>N} U_{\varepsilon,s}\varphi_s\| < \frac{\delta}{3}, \tag{8}$$

where the second inequality is satisfied uniformly in ε. The last statement follows because (6) is satisfied uniformly in ε and the boundary conditions are independent of ε. Then, using (8), we have

$$\|U_\varepsilon - U\| = \|\sum_s U_{\varepsilon,s}\varphi_s - \sum_s U_s\varphi_s\| < \tfrac{2}{3}\delta + \|\sum_{|s|\le N}(U_{\varepsilon,s}-U_s)\varphi_s\|$$

and by choosing a sufficiently small ε we can, by (7), obtain the inequality $\|U_\varepsilon - U\| < \delta$.

The necessity of having (7) satisfied for regular degeneration is rather evident.

§2. Truncation of the System

Let us now take (L_ε) to be the equation

$$L_\varepsilon U_\varepsilon \equiv \varepsilon \Delta U_\varepsilon + \mathbf{K}U_\varepsilon = 0, \tag{9}$$

where **K** is either the Cauchy-Riemann operator

$$KU = \begin{pmatrix} \dfrac{\partial u}{\partial x} + \dfrac{\partial v}{\partial y} \\[2ex] \dfrac{\partial v}{\partial x} - \dfrac{\partial u}{\partial y} \end{pmatrix}, \tag{10}$$

or the split hyperbolic operator obtained from (3) when $u=v=1$. As we noticed, a preliminary consideration of this system lets us emphasize the characteristics of the complete system (1)–(2). In this connection we shall first be interested in the negative result presented below: the lack of regular degeneracy for the operator **K** in (10). However, we begin by investigating a simpler hyperbolic case.

Thus, the operators **M** and **A** in (L_{ε}) will now be the closures of the operations D_y^2 and D_y, considered on smooth periodic functions of y. The operator **B** will be multiplication by a constant. Setting $U_{\varepsilon}=(u_{\varepsilon}, v_{\varepsilon})$, we adjoin to (9) the boundary conditions on x:

$$u_{\varepsilon}|_{x=0}=v_1(y), \qquad u_{\varepsilon}|_{x=b}=v_2(y), \tag{Γ_1}$$
$$v_{\varepsilon}|_{x=0}=v_3(y), \qquad v_{\varepsilon}|_{x=b}=v_4(y) \tag{Γ_2}$$

and consider together with (9) the unperturbed equation

$$KU = 0. \tag{11}$$

In the hyperbolic case (**K**U defined by (3) with $u=v=1$) the system (9) splits into two independent equations: an equation for u_{ε}:

$$\varepsilon \Delta u_{\varepsilon} + \frac{\partial u_{\varepsilon}}{\partial x} + \frac{\partial y_{\varepsilon}}{\partial y} = 0 \tag{12}$$

and a similar equation for v_{ε}. To each equation we must adjoin (Γ_1) or (Γ_2). Together with (12) we need to consider the unperturbed equation for u:

$$\frac{\partial u}{\partial x} + \frac{\partial u}{\partial y} = 0. \tag{13}$$

If we give $u_{\varepsilon}(x, y)$ the standard representation

$$u_{\varepsilon}(x, y) = \sum_s u_{\varepsilon, s}(x) e^{isy}, \qquad s \in \mathcal{S}, \tag{14}$$

and give $v_1(y)$ and $v_2(y)$ corresponding representations, we see that application of our usual method leads us to consider the infinite chain of equations

$$(\varepsilon D^2 + D - \varepsilon s^2 + is)u_{\varepsilon,s} = 0, \qquad D \equiv D_x, \tag{15}$$

to each of which we adjoin the conditions

$$u_{\varepsilon,s}|_{x=0} = v_{1,s}, \qquad u_{\varepsilon,s}|_{x=b} = v_{2,s} \tag{16}$$

and at the same time the chain of equations

$$(D+is)u_s = 0, \tag{17}$$

corresponding to (13).

To investigate the chain of equations (15), we set, for simplicity, $v_{1,s}=0$, $v_{2,s}=v_s$ in (16); this, as is easily seen, does not affect the qualitative nature of the picture. Then we obtain the expression

$$u_{\varepsilon,1} = v_s \frac{e^{\xi_1 x} - e^{\xi_2 x}}{e^{\xi_1 b} - e^{\xi_2 b}}, \tag{18}$$

for $u_{\varepsilon,s}$, where ξ_1 and ξ_2 are given by the formula

$$\xi_{1,2}(s) = -\frac{1}{2\varepsilon}(1 \pm \sqrt{1 + 4\varepsilon^2 s^2 - 4i\varepsilon s}). \tag{19}$$

For arbitrary $\varepsilon > 0$ and $s \in \mathscr{S}$ the values of the radicand in (19) lie on the parabola $4\zeta = 4 + \eta^2$ in the complex $\zeta + i\eta$ plane. It follows that the denominator in (18) is never zero and that the coefficient of v_s is uniformly bounded for $x \in [0,b]$, $\varepsilon > 0$, $s \in \mathscr{S}$. Consequently all the equations (15) have unique solutions for the given boundary conditions, and the solutions satisfy the inequality $\|u_{\varepsilon,s}\| \leq C|v_s|$ with a constant C that is independent of ε and s.

For a given s the behavior of ξ_1 and ξ_2 as $\varepsilon \to 0$ is described by

$$\xi_1 \approx -\varepsilon^{-1}, \qquad \xi_2 \approx -is. \tag{20}$$

In accordance with the general rules for choosing boundary conditions that guarantee regular degeneracy for ordinary differential equations (see [3]), we adjoin to the unperturbed equation (17) the condition $u_s(b) = v_s$, which yields

$$u_s(x) = v_s e^{is(b-x)}. \tag{21}$$

It follows immediately from (20) and (21) that for each given s

$$\|u_{\varepsilon,s} - u_s\| \to 0 \quad \text{as} \quad \varepsilon \to 0.$$

On the basis of the results of §1, the preceding analysis lets us state the following theorem.

Theorem 3. *Under our hypotheses on the operator* **K**, *the problem* $(\Gamma_{1,2})$ *for equation* (9) *degenerates regularly as* $\varepsilon \to 0$ *to the Cauchy problem for* (11) *with the conditions* $U|_{x=b} = v$, $v = (v_1, v_2)$.

Turning to the case when the operator **K** in (9) has the form (10), and using a representation of the form (14) for u and v, we obtain a chain of ordinary differential equations of the form

$$
\begin{aligned}
\varepsilon D^2 u_\varepsilon + D u_\varepsilon - \varepsilon s^2 u_\varepsilon + i s v_\varepsilon = 0, \\
\varepsilon D^2 v_\varepsilon + D v_\varepsilon - \varepsilon s^2 v_\varepsilon + i s u_\varepsilon = 0
\end{aligned}
\tag{22}
$$

(we do not show the dependence of u_ε and v_ε on s explicitly) and the corresponding chain of boundary conditions generated by $(\Gamma_{1,2})$. We may suppose that $s \neq 0$ (for $s = 0$ we obtain the decomposable case discussed above). After eliminating v_ε:

$$
v_\varepsilon = i s^{-1} (\varepsilon D^2 u_\varepsilon + D u_\varepsilon - \varepsilon s^2 u_\varepsilon),
$$

we obtain a fourth-order equation for determining u_ε. Write the corresponding characteristic equation $[\varepsilon^2(\xi^2 - s^2) + 2\varepsilon\xi + 1](\xi^2 - s^2) = 0$. Its roots are of the form $\xi_{1,2} = \pm s$, $\xi_{3,4} = -\varepsilon^{-1} \pm s$.

Now to carry out all the elementary calculations that are needed to establish the behavior of the family of solutions of the system (22) as they depend on $s \in S$, $\varepsilon > 0$, is rather tedious, and it is more convenient to use the results of [7].

Those results let us say that the behavior of $u_{\varepsilon, s}$ is completely determined by the properties of the determinant

$$
\begin{vmatrix}
1 & 1 & 1 & 1 \\
s & -s & p & q \\
e^{sb} & e^{-sb} & e^{\xi_3 b} & e^{\xi_4 b} \\
s e^{sb} & -s e^{-sb} & p e^{\xi_3 b} & q e^{\xi_4 b}
\end{vmatrix},
\qquad
\begin{aligned}
p &= \xi_3 + \varepsilon(\xi_3^2 - s^2), \\
q &= \xi_4 + \varepsilon(\xi_4^2 - s^2)
\end{aligned}
$$

(we have put in the values of $\xi_{1,2}$), which is the determinant of the system of equations that determine the constants c_k in the representation $u = \sum c_k \exp(\xi_k x)$, arising from the boundary conditions. The properties of this determinant establish the behavior of the roots $\xi_{k,s}$, $k = 1, 2, 3, 4, s \in \mathcal{S}$, and the choice of the boundary conditions. In applying the results of [7] we must stipulate that the parameters ε and s that appear in the boundary conditions for $u_{\varepsilon, s}(x)$ (induced by conditions (Γ_2) for v_ε) do not change the structural properties of the determinant that are used in [7]; this is easily checked. Thus we may state the following theorem.

Theorem 4. *For each given* ε *the problem* $(\Gamma_{1,2})$ *for the system* (9) *is correctly posed. At the same time, when* $\varepsilon \to 0$ *there does not exist an inequality* *that is uniform in ε for the solutions of* (22) *that correspond to the first inequality* (5).

In fact, it follows from the results of [7] that there is an inequality of the form $\|u_{\varepsilon,s}\| \leq C \sum |v^i_{k,s}|$, uniform in ε and s, for the solutions of (22) (i.e., for the solutions of the corresponding fourth-order equations) for our boundary conditions (two conditions for $x=0$ and two for $x=b$), if and only if there are constants M_1 and M_2 such that for all $s \in \mathcal{S}$, $1 \geq \varepsilon > 0$, the roots of the characteristic equation satisfy (with an appropriate numbering)

$$\xi_{(1),s}, \xi_{(2),s} \geq -M_1, \quad \xi_{(3),s}, \xi_{(4),s} \leq M_2. \tag{23}$$

Conditions (23) are evidently satisfied in our case for every given $\varepsilon > 0$, but at the same time the first condition (23) fails for every M_1 and sufficiently small $\varepsilon > 0$.

The proposition on the well-posed character of problem $(\Gamma_{1,2})$ for a given $\varepsilon > 0$ does not exclude the possibility of insolvability for certain values of ε (that play the role of a spectral parameter), and certain values of $s \in \mathcal{S}$, in the boundary value problem for (22), because of the zeros of the fundamental determinant.

§3. The Complete System

We turn to the consideration of the complete system $(1)-(2)$. In the first instance we consider the case of an operator $\mathbf{K}$ of the form (10). With the previous notation we can write the chain of equations obtained for (1) (2) in the form

$$\varepsilon D^2 u_\varepsilon + D u_\varepsilon - \varepsilon s^2 u_\varepsilon + i s v_\varepsilon + D p_\varepsilon = 0,$$
$$\varepsilon D^2 v_\varepsilon + D v_\varepsilon - \varepsilon s^2 v_\varepsilon - i s u_\varepsilon + i s p_\varepsilon = 0, \tag{24}$$
$$D u_\varepsilon + i s v_\varepsilon = 0$$

(again we do not show the dependence of u_ε, v_ε and p_ε on s). We must adjoin to (24) the boundary conditions $(\Gamma_{1,2})$. To simplify the calculations we set

$$u_\varepsilon|_{x=0} = 0, \quad u_\varepsilon|_{x=b} = v(y), \quad v_\varepsilon|_{x=0} = v_\varepsilon|_{x=b} = 0. \tag{25}$$

The case $s=0$ again requires separate discussion, which this time has to be done more carefully. The presence of the third equation (24) generates, as usual, the necessity for some consistency conditions, which in the present case lead to the requirement $v_0 = 0$ in the expansion $v(y) = \sum_s v_s e^{isy}$. The value $p_{0,\varepsilon} = \text{const}$ remains arbitrary. The corresponding characteristic equation has the form

$$[\varepsilon(\xi^2 - s^2) + \xi](\xi^2 - s^2) = 0$$

and has roots

$$\xi_{1,2}=\pm s, \quad \xi_{3,4}=-\frac{1}{2\varepsilon}(1\pm\sqrt{1+4\varepsilon^2 s^2}). \tag{26}$$

The fundamental determinant (of the system of equations for determining the v_s that come from the boundary conditions) has the form

$$D(s,\varepsilon)=\begin{vmatrix} 1 & 1 & 1 & 1 \\ s & -s & \xi_3 & \xi_4 \\ e^{sb} & e^{-sb} & e^{\xi_3 b} & e^{\xi_4 b} \\ s e^{sb} & -s e^{-sb} & \xi_2 e^{\xi_3 b} & \xi_4 e^{\xi_4 b} \end{vmatrix}. \tag{27}$$

We will show that for sufficiently small $\varepsilon>0$, $s\neq0$ (which implies $|s|\geq1$) we have $D(s,\varepsilon)\neq0$, from which it follows that the equations (24) are uniquely solvable. It is convenient to suppose that $b\geq1$. If we expand the determinant by the first row and observe that the sum of the negative terms has the form (we assume $s\geq1$) $d_1=2s(\xi_3-\xi_4)(1+e^{-b/\varepsilon})$, and that among the positive terms there are terms of the form $d_2=(\xi_3+s)(\xi_4-s)e^{(\xi_4+s)b}$, we obtain

$$d_1+d_2>\frac{s}{\varepsilon}[(\sqrt{1+4\varepsilon^2 s^2}-2\varepsilon s)e^s-2\sqrt{1+4\varepsilon^2 s^2}(1+e^{-b/\varepsilon})],$$

which establishes our proposition for $s\geq1$. It remains only to notice that the determinant changes sign under the substitution of $-s$ for s.

At the same time the distribution of the signs of the roots ξ_k (two positive and two negative for all $s\neq0$, $\varepsilon>0$) guarantees, according to [7], that the first inequality (6) is satisfied under conditions (25), uniformly in s and ε.

Turning to the unperturbed system (obtained from (24) for $\varepsilon=0$) and again dropping v and p, we adjoin to the resulting third-order equation for u_s: $D(D^2-s^2)u_s=0$ the conditions

$$u_s|_0=0, \quad u_s|_b=v_s, \quad u_s'|_b=0. \tag{28}$$

The determinant needed for finding u_ε is obtained from (27) with $\xi_4=0$ by striking out the second row and third column, and is different from zero for $s\neq0$.

Carrying out the discussion, we conclude that the fundamental proposition is satisfied for the system (1)–(2) and the unperturbed system. It remains only to verify that, for each given s, problem (25) degenerates regularly to (26) as $\varepsilon\to0$. This can be established by using the results of (3), or directly. Hence we have the following theorem.

Theorem 5. *The solution of the system* $(1)-(2)$, *where* **K** *is the Cauchy-Riemann operator, considered under the boundary conditions* (25), *degenerates regularly as* $\varepsilon \to 0$ *to a solution of the unperturbed problem, considered under the conditions* $u|_0 = 0$, $u|_b = v$, $v|_b = 0$.

When **K** is a split hyperbolic operator there is a similar theorem, whose proof is considerably simpler.

Appendix 2
Further Developments

(Professor Dezin asked me to summarize five papers that appeared after the publication of his book. The summaries are given in chronological order. Papers cited in this appendix are listed on p. 157. - Translator.)

Yunusov, M.: Operator equations with a small parameter and nonlocal boundary conditions. Differentsialnye Uravneniya *17* (1981), no. 1, 172–181; English translation in Differential Equations *17* (1981), no. 1, 121–127.

This is a continuation of Appendix 1. Under the fundamental hypothesis of §1 (p. 147), Yunusov discusses the family of operators $L_\varepsilon u = \dfrac{du}{dt} - Au = f(t)$ with boundary conditions

$$\mu u|_{t=0} - u|_{t=b} = g,$$

where A is a Π-operator. He shows that the problem degenerates regularly to the unperturbed problem L_0 if and only if $\|u_{s,\varepsilon} - u_{s,0}\| \to 0$ ($\varepsilon \to 0$) for each $s \in S$; also $u_\varepsilon = u_0 + v_{\varepsilon,\mu} + z_{\varepsilon,\mu}$, where $\|z_\varepsilon\| \to 0$ uniformly in $t \in [\delta, b-\delta]$ and v is a boundary layer function ($\|v_\varepsilon(t)\| \to 0$ uniformly for $t \in [\delta, b-\delta]$, $\|v_\varepsilon(t)\| \leq C$ for $a \leq t \leq b$. The following illustrations are discussed: $A = D_x^2 + D_y^4 \pm d$; $A = D_x^2 - D_y^2 - d$ ($d > 0$, not an integer); $A = D_x^2 - D_y^2 + d$, $d \neq 0$, not an integer.

Dezin, A.A.: On overdetermined boundary value problems. Sibirsk. Mat. Zh. *24* (1983), no. 5, 43–47; English translation in Siberian Math. J. *24* (1983), no. 5, 681–685.

The author shows that for the operators A of §2.1 the Cauchy problem ($L-C$ problem) $L_u = \dfrac{du}{dt} - Au = f$ with $u|_{t=0} = 0$ is proper if and only if there is a number $M < \infty$ such that $\operatorname{Re} A(s) < M$ for all $s \in S$. In this connection he conjectures that a *necessary* condition for there to exist a system of boundary conditions Γ such that L^{-1} is bounded is that the periodic $L-C$ problem is proper. This would follow from the conjecture that if the $L-C$ problem is not proper then L_0^{-1} (corresponding to the case $A_\Gamma = A_0$) is unbounded. Several illustrative examples are presented.

This paper supplements Chapter V, particularly §2, and refers to [A1].

Kornienko, V.V.: On the spectra of irregular operators. Differentsialnye
Uravneniya *21* (1985), no. 1, 65–77; English translation in Differential Equations *21* (1985), no. 1, 52–61.

This paper is connected with Chapter VI, §1. It uses Dezin's methods
[26] to study the problem

$$Lu = a(t) D_t^2 u - A(t) u = f,$$

$$a(t) = \begin{cases} a_1, & 0 \leq t < h, \\ a_2, & h \leq t \leq b, \end{cases} \qquad A(t) = \begin{cases} A_1, & 0 \leq t < h, \\ A_2, & h \leq t \leq b, \end{cases}$$

where the A_k are differential operators of the form (1), §2, Chapter II,
assumed to be Π-operators. By following the proof of a theorem of Dezin
[A2] the author shows that if $f = f(x,t) = \sum_s f_s(t) e^{is \cdot x}$, then the general
solution of the problem with given linear homogeneous boundary conditions on t and x exists and is unique for every $f \in H$, if and only if the
ordinary differential equation

$$L_s u_s(t) \equiv a(t) D_t^2 u_s(t) - A(s,t) u_s(t) = f_s(t),$$

$$A(s,t) = A_1(s), \quad 0 \leq t < h; \qquad A_2(s), \quad h \leq t \leq b,$$

has a unique solution for each $f_s(t) \in H_t$ and $\|u_s\| \leq c \|f_s\|_t$, where c is inde-
pendent of s. The author studies the problem for the cases $a(t) \equiv 1$ and
$A(t) \equiv A$. He obtains quite detailed information about the spectrum and the
eigenfunctions, and finds cases of weak or strong irregularity.

Dezin, A.A.: Spectral characteristics of general boundary value problems
for the operator D^2. Mat. Zametki *37* (1985), no. 1–2, 249–256; English
translation in Math. Notes *37* (1985), no. 1–2, 142–146.

Let L be a regular ordinary differential equation of order n on a finite
interval. The spectrum depends on the boundary conditions via n^2 essential
parameters (§§2, 3, Chapter III) but the general theory does not say how to
determine these parameters. Here, when $n = 2$, the parameters are chosen to
allow a study, more detailed than usual, of how the spectrum depends on
the boundary conditions. The essential idea is a new representation of the
Green function. The problem reduces to an equation (L): $L_\lambda(\Gamma) \equiv D_t^2 u - \lambda u$
$= f$, and this in turn reduces the problem to a study of an equation $\Delta(\alpha)$
$= \det A(\alpha)$. The author describes the spectrum, under various hypotheses
about the boundary conditions Γ, for the Dirichlet and Neumann problems,
the periodic Cauchy problem, and the perturbed Neumann problem.

This paper could be read after Chapter VI, §1.

Romanko, V.K.: Nonlocal boundary value problems for certain systems
of equations. Mat. Zametki *37* (1985), no. 5–6, 727–733; English translation
in Math. Notes *37* (1985), no. 5–6, 400–404.

On a finite interval, consider systems

$$Lu(x, t) = D_t u - AP(D_x) u = f(x, t),$$

where u and f are 2-vectors, A is a square matrix, and P is an operator as in (1), §2, Chapter II. The author obtains a condition for L to be proper, and a condition on A under which there is no t-boundary condition making L proper.

References for Appendix 2

A1. Dezin, A.A.: Degeneracy of operator equations. Mat. Sb. *115* (1981), no. 3, 323–336; English translation in Math. USSR Sb. *43* (1982), no. 3, 287–298

A2. Dezin, A.A.: On weak and strong irregularity. Differentsialnye Uravneniya *17* (1981), no. 10, 1851–1858; English transl. in Differential Equations *17* (1981), no. 10, 1160–1165

References

I. Books

1. Ahiezer, N.I., Glazman, I.M.: Theory of linear operators in Hilbert space. Ungar, New York, vol. 1, 1961; vol. 2, 1963
2. Berezanskiĭ, Yu.M.: Expansions in eigenfunctions of selfadjoint operators. American Mathematical Society, Providence, R.I. 1968
3. Besov, O.V., Ilin, V.P., Nikolskiĭ, S.M.: Integral representations of functions, and embedding theorems (in Russian). Nauka, Moscow 1975
4. Cassels, J.W.S.: An Introduction to Diophantine approximation. Cambridge University Press 1957
5. Dunford, N., Schwartz, J.T.: Linear operators, vol. 1. Interscience, New York 1958
6. Dunford, N., Schwartz, J.T.: ibid., vol. 2, 1963
7. Gårding, L.: Cauchy's problem for hyperbolic equations. Lecture Notes, University of Chicago 1957
8. Halmos, P.R.: Finite dimensional vector spaces. Princeton University Press 1942
9. Hörmander, L.: Linear partial differential equations. Springer-Verlag, Berlin-Heidelberg 1963
10. Kantorovich, L.V., Akilov, G.P.: Funktionalanalysis in normierten Räumen. Harri Deutsch, Thun-Frankfurt am Main 1978
11. Kochin, N.E., Kibel, I.A., Roze, N.V.: Theoretical hydromechanics (in Russian). Fizmatgiz, Moscow 1963
12. Kolmogorov, A.N., Fomin, S.V.: Eléments de la théorie des fonctions et de l'analyse fonctionelle. Mir, Moscow 1974
13. Kreĭn, S.G.: Linear differential equations in Banach space. American Mathematical Society, Providence, R.I. 1971
14. Lions, J.L.: Equations différentielles-operationelles et problèmes aux limites. Springer-Verlag, Berlin 1961
15. Lorch, E.R.: Spectral theory. Oxford University Press, New York 1962
16. Lyusternik, L.A., Sobolev, V.I.: Elemente der Funktionenanalysis. Akademie-Verlag, Berlin 1965
17. Nagumo, M.: Lectures on the contemporary theory of partial differential equations (in Russian; translated from the Japanese). Mir, Moscow 1967
18. Naĭmark, M.A.: Linear differential operators, Part I, 1967; Part II, 1968. Ungar, New York
19. Nikolskiĭ, S.M.: Approximation of functions of several variables, and embedding theorems (in Russian). Nauka, Moscow 1960
20. Sobolev, S.L.: Some applications of functional analysis in mathematical physics (in Russian). Leningrad State University Press 1950
21. Yosida, K.: Functional analysis, 3rd ed. Springer-Verlag, Berlin-Heidelberg-New York 1971

II. Papers

22. Dezin, A.A.: Boundary value problems for some symmetric linear first-order systems. Mat. Sb. *49* (91) (1959), no. 3, 459–484

23. Dezin, A.A.: Existence theorems and the uniqueness of solutions of boundary value problems for partial differential equations in function spaces. Uspehi Mat. Nauk *14* (1959), no. 3, 21–73

24. Dezin, A.A.: The simplest solvable extensions for ultrahyperbolic and pseudoparabolic operators. Dokl. Akad. Nauk SSSR *148* (1963), no. 5, 1013–1016; English translation in Soviet Math. Dokl. 4 (1963), no. 1, 208–211

25. Dezin, A.A.: Operators with a first derivative with respect to "time", and nonlocal boundary conditions. Izv. Akad. Nauk SSSR, Ser. Mat. *31* (1967), no. 1, 61–86

26. Dezin, A.A.: On some systems of equations containing a small parameter. Mat. Sb. *111* (153) (1980), no. 3, 323–333; English translation in Math. USSR Sb. *39* (1981), no. 3, 289–298

27. Dikopolov, G.V., Shilov, G.E.: On correctly posed problems for partial differential equations in a half space. Izv. Akad. Nauk SSSR Ser. Mat. *24* (1960), no. 4, 369–380

28. Dubinsky, Yu.A.: On some differential-operator equations of arbitrary order. Mat. Sb. *90* (132) (1973), no. 1, 3–22; English translation in Math. USSR Sb. *19*, nos. 1–4 (1973), 1–21

29. Friedrichs, K.O.: The identity of weak and strong extensions of differential operators. Trans. Amer. Math. Soc. *55* (1944), 132–151

30. Grothendieck, A.: La théorie de Fredholm. Bull. Soc. Math. France *84* (1956), 319–384

31. Hörmander, L.: On the theory of general partial differential operators. Acta Math. *94* (1955), 161–248

32. Hörmander, L.: Differential operators of principal type. Math. Ann. *140* (1960), 169–173

33. Hörmander, L.: Weak and strong extensions of differential operators. Comm. Pure Appl. Math. *14* (1961), 371–379

34. Lax, P.D., Phillips, R.S.: Local boundary conditions for dissipative symmetric linear differential operators. Comm. Pure Appl. Math. *13* (1960), 427–455

35. Leray, J.: Valeurs propres et vecteurs propres d'un endomorphisme complètement continu d'un espace vectoriel à voisinages connexes. Acta Sci. Math. Szeged. *12*, part B, 177–186 (1950)

36. Mamyan, A.H.: The construction of solvable extensions in a parallelepiped for linear partial differential operators. Differentsialnye Uravneniya *6* (1970), no. 2, 358–370; English transl. in Differential Equations *6* (1972), no. 2, 278–287

37. Mihaĭlov, V.P.: On Riesz bases in $\mathscr{L}^2(0,1)$. Dokl. Akad. Nauk SSSR *144* (1962), no. 5, 981–984; English translation in Soviet Math. Dokl. *3* (1962), no. 3, 851–855

38. Romanko, V.K.: On the theory of operators of the form $\dfrac{d^m}{dt^m} - A$. Differentsialnye Uravneniya *3* (1967), no. 11, 1957–1970; English translation in Differential Equations *3* (1967), no. 11, 1018–1025

39. Romanko, V.K.: Boundary value problems for a class of differential operators. Differentsialnye Uravneniya *10* (1974), no. 1, 117–131; English translation in Differential Equations *10* (1975), no. 1, 84–94

40. Romanko, V.K.: Single-valued solvability of boundary value problems for some differential-operator equations. Differentsialnye Uravneniya *13* (1977), no. 2, 324–335; English translation in Differential Equations *13* (1975), no. 2, 225–233

41. Romanko, V.K.: On boundary value problems for differential equations that are not solvable for the highest derivative. Dokl. Akad. Nauk SSSR *235* (1977), no. 5, 1030–1033; English translation in Soviet Math. Dokl. *18* (1977), no. 4, 1101–1105

42. Sebastião e Silva, J.: Sur le calcul symbolique d'opérateurs permutables, à spectre vide ou non borné. Ann. Mat. Pura Appl. (4) *58* (1952), 219–275

43. Sebastião e Silva, J.: Su certe classi di spazi localmente convessi importanti per le applicazioni, Rend. Mat. e Appl (5) *14* (1955), 388–410

44. Vishik, M.I.: On general boundary value problems for elliptic differential equations (in Russian). Trudy Moskov. Mat. Soc. *1* (1952), 187–246

Index

Index of Symbols

V. G. Maz'ja

Sobolev Spaces

Translated from the Russian by
T. O. Šapošnikova

Springer Series in Soviet Mathematics

1985. 25 figures. XIX, 486 pages.
ISBN 3-540-13589-8

The various aspects of the theory of Sobolev spaces are considered in this monograph. Particular attention is paid to the so-called imbedding theorems which were originally established by S. L. Sobolev in the 1930s and proved to play a significant role in functional analysis and in the treatment of linear and nonlinear partial differential equations.
A large part ot the basic results achieved in Sobolev space theory was, up till now, scattered in various journals. This book draws these results together and provides a unified treatment of the theory. An account of new results is also given. Many examples are included to help achieve a better understanding of the subject. Student who would like to acquaint themselves with this important field of mathematics as well as researchers will find this book a useful basis.

Springer-Verlag
Berlin Heidelberg New York
London Paris Tokyo

O. I. Bogoyavlensky

Methods in the Qualitative Theory of Dynamical Systems in Astrophysics and Gas Dynamics

Translated from the Russian by D. Gokhman

Springer Series in Soviet Mathematics

1985. 40 figures. IX, 301 pages. ISBN 3-540-13614-2

Contents: Methods of Qualitative Analysis of Multi-Dimensional Dynamical Systems. – Qualitative Theory of Homogeneous Cosmological Models Without the Motion of Matter. – Qualitative Theory of Homogeneous Cosmological Models with the Motion of Matter and Electromagnetic Fields. – Self-Similar Spherically Symmetric Solutions for the General Theory of Relativity. – Self-Similar Motion of Self-Gravitating Gas in Stars. – Self-Similar Rotation of an Ideal Gas. – The Dynamics of a Gaseous Ellipsoid. – The Dynamics of Perturbations of the Periodic Toda Lattice. – Bibliography.

This book is devoted to the qualitative analysis of multi-dimensional dynamical systems. The author develops applications to several important problems of cosmology, theoretical astrophysics and gas dynamics. This new method of maximally non-degenerate compactification of a dynamical system is used to analyse models of various cosmological events including models of supernovae and accretion of matter onto black holes, models of oscillations of rotating gas nebulae and various vorticities in gases and fluids.
This book provides a thorough introduction to this important field of mathematics and astrophysics and equips the reader well in dealing with a wide range of concrete problems.

Springer-Verlag
Berlin Heidelberg New York
London Paris Tokyo